Younger Today

The Cell Solution To Youthful Aging And Improved Health

Vincent C. Giampapa, MD
Carol Alt

16
EasyRead Large

RHYW

Copyright Page from the Original Book

The information contained in this book is based upon the research and personal and professional experiences of the author. It is not intended as a substitute for consulting with your physician or other healthcare provider. Any attempt to diagnose and treat an illness should be done under the direction of a healthcare professional.

The publisher does not advocate the use of any particular healthcare protocol but believes the information in this book should be available to the public. The publisher and author are not responsible for any adverse effects or consequences resulting from the use of the suggestions, preparations, or procedures discussed in this book. Should the reader have any questions concerning the appropriateness of any procedures or preparation mentioned, the author and the publisher strongly suggest consulting a professional healthcare advisor.

Basic Health Publications, Inc.
28812 Top of the World Drive
Laguna Beach, CA 92651
949-715-7327 • www.basichealthpub.com

Library of Congress Cataloging-in-Publication Data is available through the Library of Congress.

ISBN: 978-1-59120-716-0

Editor: Emily Hubbell
Project editor: Carol Killman Rosenberg
Typesetting/Book design: Gary A. Rosenberg
Cover design: MUDHAUS
Cover photography: Jimmy Bruch

Printed in the United States of America

1 0 9 8 7 6 5 4 3

TABLE OF CONTENTS

Introduction

What does your "ten years from now" look like?

No matter who you are, where you are in life, or what makes you tick, one thing ties us all together: we want to live a long, healthy life. The problem is that our world's population is aging at rocket speed—so quickly that in three short decades, one in five Americans you know will be older than sixty-five. Our healthcare system won't be able to keep all those aging people healthy, but that's only the half of it. Stuck with low energy, fatigue, damaged immune systems, and other health issues, the people we know and love will have less energy to do the things they've dreamed of doing. They'll be spending less time living life and more time in the doctor's office.

But what if we told you there's a way to bypass all that and extend your youthfulness and health into your sixties and seventies? What if we could look and feel younger today? We can.

The secret to youthful aging and a healthier life isn't a potion or a lotion. It isn't a gimmick or a chemical with a multisyllabic name and endless side effects. It isn't a surgical procedure that requires months of recovery, either. The secret to a healthier, more youthful life is right in your own body—in fact, it's right in those genes you've always been told you're stuck with. The secret to youthful aging is the Cell

Solution. With the right lifestyle changes, you can reprogram your genes to be healthy, robust defenders of your youth. The healthier your cells, the more youthful you are, from the inside out. It's time to make your "golden years" golden again.

Why are a model and a surgeon coming together to write this book? Because we've both been in your shoes. We know what it's like to feel less than healthy. We know what it's like to be worried about aging and the ailments that come with it. We know what it's like to see the people we love get older too early in their lives. But we also know that the secret to mitigating these issues is nurturing our genes through diet, exercise, stress-coping strategies, and most importantly, all-natural phytonutrient complexes that slow aging at its roots. We're excited to help you discover your youthful today.

***

I've always been the guy who asks too many questions. During my first few years as a plastic surgeon, I watched countless patients be wheeled in and out of the operating room. Sometimes they recovered youthfully; sometimes they didn't. I remember asking my supervisor what else we could do to make sure our patients were healthier going into surgery, but my question fell on deaf ears. Later, in my own practice, I witnessed how my cosmetic procedures reversed the telltale effects of aging in my patients. But reversing aging wasn't enough; I wanted to get to the root of why we age prematurely. Why did some of those surgery patients have a more youthful recovery? How could I help my patients look *and* feel younger? These questions compelled me to explore the origin of human aging. I became immersed in the search.

Fast-forward a few years. I was busy founding and growing the American Academy of Anti-Aging Medicine and still researching the roots of aging when something profound happened. We all have moments that make us stop, think, and change course. For me, that moment was my father's death at the hands of the same rare cancer that killed his brother. I was devastated by the loss and deeply concerned. Would the same thing happen to me? Would my kids develop the same cancer that killed my father and uncle? If you carry a gene for a condition, is it inevitable that you'll succumb to it? Determined to find out, I shifted focus and pointed my anti-aging research at the cellular level. It's been my passion ever since.

Here's what I've discovered: We inherit genetic tendencies, not certainties. About 30 percent of our genetics is set in stone, but when you do the math, that leaves a big piece of the pie that we can control. Through the Cell Solution lifestyle, you can make the most of that 70 percent so that your window of health, youthfulness, and well-being goes on for longer than ever before. Information can change your habits, your habits can change your genes, and your genes can change your life.

Every day, I see cellular health transform my patients, friends, and family. The change is immediate, and it's profound. It's happened to me, it's happened to Carol Alt, and with the right strategies, it will happen to you.

***

Don't let the word "supermodel" fool you. I haven't always been the youthful, happy, energetic Carol Alt you see today. Back in my thirties, I was the picture of poor health. I was starving myself to stay model-thin, yet every time I stepped on the scale, the number jumped higher. A baked potato for lunch was enough to make me gain a pound! While on shoots, my energy was so low that every outfit change felt like a huge ordeal. I was irritable and tired, and everyone around me noticed. And don't even get me started on the medications. I needed coffee to wake up in the morning, an endless supply of Tums to get me through my day, and Nyquil to fall asleep at night. I was so plagued by headaches that I kept aspirin in every pocket. Meanwhile, my sinus infections lingered and lingered, and my allergies got progressively worse. I had sagging skin, noticeable wrinkles, and none of that youthful glow you'd expect from someone in her thirties. Unhealthy is an understatement: my body was a complete train wreck, and my weight was derailing right along with it.

You'd think that after enduring this for a few months, I would've heard my body screaming out for help.

But I lived this way for years, until one fashion shoot gave me no choice but to see just how much my habits had aged me. Picture this: I'm in my early thirties, on location in Venezuela, surrounded by beautiful rainforests. I should've been basking in the prime of my life. Instead, I was miserable from head to toe. They had told me when I came on set that my body wasn't in swimsuit shape, so I was feeling ashamed and uncomfortable from the get-go—and hiding behind every rock I could find. Then I saw her: a twentysomething model moving gracefully from rock to rock and completely captivating everyone on set. I was so distraught and depressed that I ended up leaving the shoot early. Talk about a low point!

The more I thought about that younger model and my disastrous Venezuela shoot, the more I was completely baffled. If the transition from twenty to thirty had gotten me here, what would I be like at forty? Would there be any hope for me at fifty? Or, dare I say, at sixty?

That's when I knew I needed help. Thankfully, with the guidance of my doctor, I found my saving grace: raw food. The transformation was immediate. Nearly overnight, I felt my mood shift and my energy restored. My skin tightened and finally started glowing again. And did I mention how quickly the weight started coming off? With raw food, I was suddenly living a life where I could eat as much as I wanted. I felt full and satisfied, and somehow didn't have to worry about gaining a pound. For a perpetual dieter, there's nothing more liberating than actually eating. Can you blame me for never looking back?

My lifestyle change started with food, but since everything in our body is connected like dominoes, other parts of my life quickly followed suit. The only difference is that now, instead of all those dominos falling down, they work together to maintain the healthy foundation I've given myself. It's amazing how much better I am at managing stress and how much less anxious I am when life gets hectic, on set or otherwise. I'm more energetic when I'm awake and able to sleep soundly when my body needs it. I tell people that my body is like an electric golf cart. When I put my foot on the gas, it goes, and when I put my foot on the brake, it stops. And I look better, too! People stop me every day and tell me it looks like I've just been on a long, relaxing vacation. My skin is glowing and firm, my body is energized, and I'm ready to seize the day just as I did in my twenties—minus the Tums and aspirin, of course.

What does eating raw have to do with cell health? Everything. Through our approach, you'll give your cells the phytonutrients that help them function best. We're really doing the same thing a raw diet does: ditching the chemicals, processed materials, and other junk that triggers the kind of sad situation I was in seventeen years ago. Take it a step further, and you'll tap into a lifestyle where your exercise and habits are in tune with what your cells actually need to stay younger. The Cell Solution is like giving your entire life a "go raw" makeover. It's a green juice diet without the cleanup!

People always tell me, "I'm not like you. This won't work for me." It baffles me every time. Not like me? What do you mean? Every person is the same under the skin. The body needs what the body needs, and we have to satisfy those needs at the cellular level.

Dr. Giampapa and I are here to help you recapture your youthfulness today. Who says we can't control aging?

HOW HEALTHY ARE YOU?

On any given day, our body communicates with us in many different ways. Not surprisingly, a lot of it has to do with how healthy we are. What signals is your body sending you on a daily basis? Just how healthy are you? If you answer "yes" to any of the following questions, your body is telling you that your health could use a boost. Think about these questions as you embark upon your healthy cell lifestyle.

1. Do you have trouble falling asleep at night? ______
2. Once you fall asleep, do you wake up repeatedly throughout the night? ______
3. Does your brain often feel like it's in a fog? ______
4. Do you feel fatigued during the day? ______
5. Do you need caffeine and/or sugar to get through your day? ______
6. Are you easily irritated? ______
7. Do you have the energy you need to do all the things you want to do? ______
8. Is your hair dull or dry? ______
9. Do you lack the vigor you had a few years ago? ______
10. Do you often need pain relievers, cold medicine, or other over-the-counter medications to get you through your day? ______

What do all these health issues have in common? They can improve if your cell health improves. Get ready to dive into your youthful today.

1

A More Youthful You

"If only I could feel young again."

How many times have you heard that one before? Worse yet, how many times have you thought it yourself?

We've all been there: You're going through your hectic day, and somewhere along the way, your energy vanishes into thin air. You're dragging all afternoon. When you finally get home from work, exhausted, you happen to glance in the mirror as you walk to the kitchen. That's a mistake. Why do you look so fatigued? Is that a wrinkle coming in? Where's that vibrancy and zeal you had a few years ago? As you turn sideways to study your figure, a familiar, nagging thought pops into your mind. "If only I could feel young again."

It's time to banish that thought once and for all. You can recapture your youthfulness, and you can do it today. It's just a matter of getting your genes working in your favor. We're here to show you how.

We're told that we're stuck with our genes from birth—through good and bad, rain and shine, thick and thin, our genes determine how youthful we look and how quickly we age. But that's simply not true. Only about 30 percent of your genetic makeup is set in stone, and what's left over is entirely in your hands. When it comes to aging, that 70 percent is your ticket to youth. With the right lifestyle and nutrients, you can reprogram your genes so that they maximize your health and slow aging. Whether you want to recapture your energy and vigor or simply want to feel better every day, a healthier life is within reach at any age. It starts with your cells, and it starts today.

HEALTHIER CELLS, HEALTHIER YOU (AT ANY AGE)

Cardiovascular health. Respiratory health. Immune health. Digestive health. If you don't know where to start, figuring out how to stay youthful is like opening Pandora's box. Our bodies have eleven systems (digestive, skeletal, immune, and so on) and even more subsystems, and somewhere along the way, we've come up with a separate set of health rules for each. How are any of us supposed to live real lives, stay young and vibrant, and juggle a mountain of diet and exercise tips? Luckily, there's an approach that has all your body's bases covered. Ditch all your health how-to's and make this your mantra: your health begins and ends with your cells.

Everything in your body is connected, from your largest organ, the skin, down to the tiny cells that compose it. Do you see that beautiful hair on your head? For that to grow, the body's systems had to come together and pull off countless processes without a hitch. To catch those eight hours of z's last night, your brain, retina receptors, muscles, hormones, and numerous other organs and body systems had to be on board. The parts of your body are in such a close relationship that the tiniest change in one area can make the rest of the "ship" change course. But that's the beauty of cell health. When you keep your body's 100 trillion building blocks healthy, your entire body's health falls into place. It's cardiovascular health, muscular health, and all those other types of body health rolled into one dynamic little package.

When you keep your cells healthy, your entire body's health falls into place. It's cardiovascular health, muscular health, and all those other types of health in one dynamic little package.

Cell health doesn't just boost every part of your body; it also transforms you at any age. It can give you more energy to fuel your daytime routine and help you finally experience that better night's sleep you've been dreaming about. It can help stabilize your metabolism so that you can get the most out of your food and exercise. It can get all your systems in synch so that they're working together in harmony. In short, healthy little cells bring big health benefits, no matter what age is on your driver's license. Cell health equals better health for the entire body, for every person, at every age. Consider it the universal healthcare regimen.

Sure, your cells build up your organs and tissues, but what does that have to do with the new wrinkle on your face or the healthy spring in your step? Everything. Healthy cells mean a healthy body, and a healthy body means a more youthful you. When your cells are in tip-top shape, they'll give you the best gift on Earth: more energetic, feel-good years and fewer days of body breakdown. But when your cells are damaged, they start to turn toxic. They pollute the other cells around them and threaten your entire body's happy, healthy rhythm. That's when premature aging and poor health show up to ruin the party.

Imagine that your body is a symphony. When the clarinet player in the third row hits a wrong note, it distracts the musician one seat over. If the note is bad enough, the next player in the row is thrown off too. Soon, the whole clarinet section loses its rhythm, the tubas can't follow along, the trumpets are a measure behind, and the conductor is trying frantically to get the whole group back in synch. Meanwhile, everyone in the audience can tell that something's not right, even from all the way in the back row. Those wrong notes are unhealthy, damaged cells.

When your body's symphony goes from harmony to complete discord, you'll feel changes everywhere, from your brain and heart right down to your toes. Yes, we all age naturally. That's okay. But what we're talking about here is premature aging, the kind of body breakdown that makes us look and feel a lot older than we should. Here's a quick rundown of some conditions associated with premature aging:

- Lower energy
- Increased body-fat levels
- Poor muscle tone
- Loss of endurance
- Loss of libido
- Impaired eyesight
- Reduced cognitive function
- Slower healing
- Wrinkles and loss of skin elasticity Think those were bad? Here's the more serious stuff:
- Heart disease
- Cancer
- Neurodegenerative diseases
- Arthritis
- Diabetes

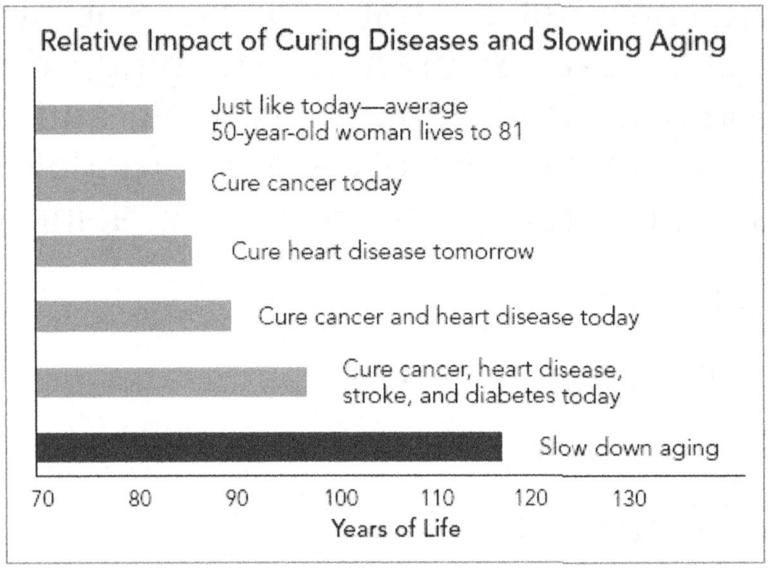

Figure 1.1. If we slow down aging, our lives get a boost in more ways than one. Slowing aging would increase the average human life span more than curing cancer, diabetes, heart disease, and stroke.

When your cells deteriorate, you're not just less healthy; you're also aging much quicker than you should! The trick to staying youthful and healthy later in your life is keeping those bad notes from taking over the whole show. That's where the Cell Solution comes in.

DNA DAMAGE DEBACLE

Before we can talk about how to keep your body's symphony in harmony, let's get to the bottom of how your cells become bad notes. It all boils down to DNA damage, which causes your cells to transition from healthy, happy, and harmonious to unhealthy, damaged, and senescent. Here's the full story behind "senescence," a fancy word for the aging and toxicity that occurs in your cells when they are badly damaged over time.

Let's start with the basics. Cells aren't just the bricks and mortar that build healthy organs, tissues, and systems in your body; they're also the home to the DNA that makes you who you are. Your genes are stored on strands of DNA called chromosomes, which turn on and off in order to make sure your body creates all the substances you need to stay alive. When your DNA's "on" and "off" switches tell the beneficial processes to kick in, your genes are on your side. You're feeling like a million bucks. When miscommunication mucks up your gene switching, aging starts to creep in. And we're not just talking wrinkles and saggy skin: bad switching can turn on genes that accelerate the onset of those diseases we just listed for you. Yikes.

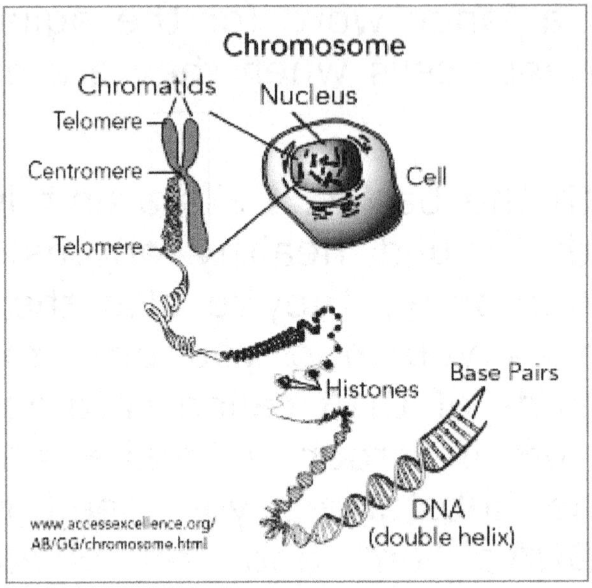

Figure 1.2. Your cells are home to your DNA.

Try thinking about it this way: in the younger you, your cells knew you needed to grow, so the genes that spur those processes switched on. As you age, your cells switch off some growth genes you no longer need and activate genes that keep your body feeling fit, energetic, and strong. And no matter your age, fight-or-flight genes kick in when your body senses danger or stress. This gene switching is your body's way of responding to whatever your environment throws your way.

In a perfect world, the genes you need are switched "on," and the ones you don't need are switched "off." DNA damage is like the wrong note coming from the clarinet player. When your DNA is injured, your genes can't operate at their best. Before long, mistakes start creeping into your gene switching. More of your helpful genes go into "off mode." The "old age and disease" gene switches can be flicked on, and some of the defenders that your body relies on to ward off health issues can shut off—right when you need them most. Talk about bad notes! When more of your genes are turned "off" and the ones left "on" are out of synch with what your body needs, you've entered the land of disharmony and premature aging—a place where none of us wants to spend much time.

How does this tie back to your cell health? If cells can't repair their damaged DNA, they become unhealthy toxic waste dumps that pollute the other cells around them. A healthy cell is a model spokesperson who improves the community around it, but an unhealthy, toxic cell shakes up the whole neighborhood and throws off the harmony just like that pesky clarinet player. When a wave of toxicity spreads throughout your cells, you're headed for a major case of the aging blues characterized by diabetes, obesity, bone pain, body fat, and—you guessed it—that dreaded saggy skin.

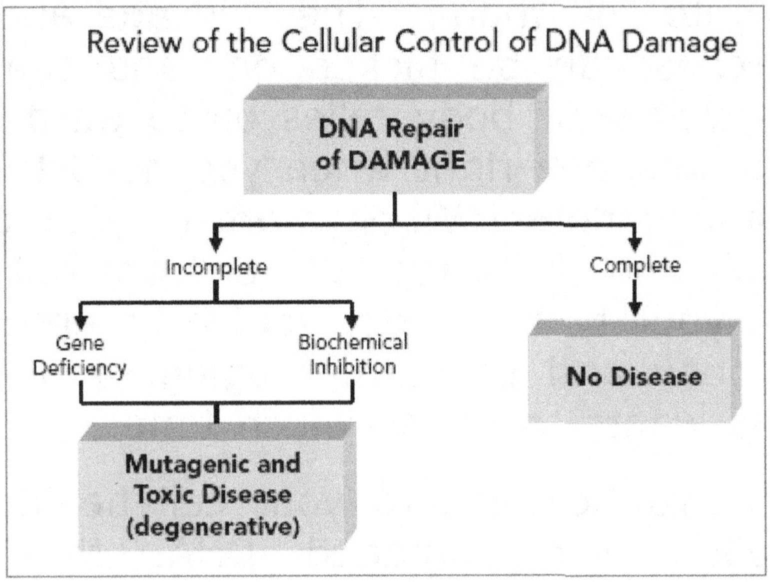

Figure 1.3. DNA damage can trigger major diseases. It's just another reason why healthy cells make a healthy you.

It all boils down to this: your genes don't seal your fate. If you can boost your cell health enough to keep your DNA strong and functioning, you can make the most of the genes you're born with. Your cells will be good, youthful neighbors who keep the whole community healthier for longer than ever before, and you'll be looking and feeling great whether you're in your twenties or your retirement years.

TOXIN TALK

How does your precious DNA get damaged in the first place? If we had to place the blame somewhere, it would be on free radicals. Here's the lowdown on these toxic visitors:

Our body isn't naturally a beautiful, squeaky-clean oasis. On any given day, we all battle free radicals; it's as much a part of us as getting energy from our food. When our bodies change food into fuel, it causes some atoms or molecules to lose an electron. This turns the atom or molecule into an unstable little free radical that tries to "steal" electrons from another cell to regain its balance. But here's the thing: when a free radical snatches away an electron, the cell it steals from becomes a new free radical. The dominos just keep falling from there.

The creation of free radicals is just a fact of life, and our bodies can fend against these toxic molecules in moderation. The problem is that these days, free radicals are a bigger part of our life than ever before—thanks in part to things like constant stress and processed food, both of which flood our bodies with these volatile molecules. With more than 10,000 highly charged free radicals damaging each cell daily, it's no wonder our DNA is being beat up and our healthy years are being cut short.

Free radicals themselves might be invisible to the eye, but you'll notice, and feel, their byproducts right away. These volatile little molecules and atoms don't just scramble the codes in our DNA; they're also some of the prime suspects in the crow's-feet, frown lines, and wrinkles you see. Free radicals come from the food we eat, the pills we pop, the air we breathe, and the other toxic elements of our environments. Are you starting to see how your actions and genes collide?

PHOTOCOPYING YOUR LIFE

We know that free radicals bombard our cells with toxicity, and that toxicity can injure our DNA and our cells. If our lifestyle causes many free radicals to enter our bodies, that damage can spread quickly. But just how do we go from having a few damaged cells to having a body full of them? A big factor is cell replication.

Our cells have a life cycle, just as we do. When they are damaged, they divide so that your body has a constant influx of worker-bee cells. These divisions are a normal part of life, but they're also a factor in a cell's death. Each time a body cell splits in two, its lifespan is shortened. After fifty or sixty divisions, the little end caps on your chromosomes, your telomeres, are so truncated that your cell can't duplicate itself anymore. This signals that it's time for your old cell to join all the other malfunctioning cells in the sky. It may sound sad, but believe us: your body is better off without all those old, unproductive cells floating around. This is just your body's way of cleaning house.

Your body also has a way of repopulating its cell workforce each time one of your cells dies off. That's where stem cells come in. Once a cell is ready to die, your adult stem cells "hear it through the grapevine" and quickly arrive at the scene to make a copy of the dying cell before it's gone for good. But here's the best part: This isn't just an exact copy of the faulty cell. Functioning stem cells can turn back the clock so that the copy they create is as healthy as the dying cell was in its pre-damage glory days. These healthy, youthful copies ensure your body functions like a well-oiled machine.

Stem cells get a bad rap these days, but the cells we're talking about here aren't found in embryos —they're found naturally in all the tissues of your body. These little guys can be activated to transform into any kind of cell in their surrounding tissue, and that's why they're so important to staying healthy and youthful. Without healthy stem cells, you can kiss youthful cells—and your youth—good-bye.

Functioning stem cells make accurate copies. In a perfect world, that would be the end of the story. But the reality is that stem cells age just as your other cells do. Until we hit thirty years old, our stem cells can fix all their DNA injuries; that means the copies they create are youthful, functioning laborers that keep our tissues and systems healthy. But starting in our thirties, our DNA becomes less adept at repairing itself. When the DNA in your stem cells is so damaged that it's not fixable, the copies your cells create are damaged, too. This floods your system with less-than-perfect worker bees. You can see where this is going.

It's a little like photocopying a poor-quality image: each time, the output gets a bit more pixilated and a bit fuzzier. That may not be a big deal when you're making a copy of a snapshot or work document, but when this reduced quality happens within your DNA—the most important information in your body—there's no underestimating the impact. Over time, these low-res copies trigger diseases and contribute to major wear and tear on your body's systems. When it gets so bad that your adult stem cells stop functioning or die off en masse, your body can experience everything from muscle mass and hormone decline to bacterial and viral infections.

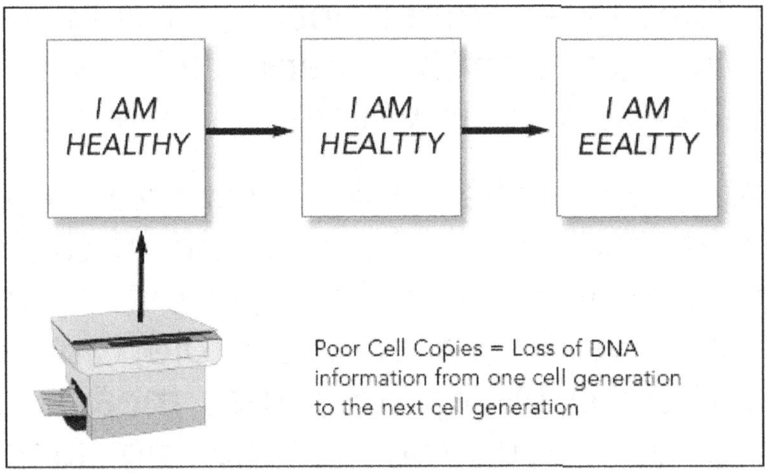

Figure 1.4. Unhealthy cells make bad copies. Bad copies can trigger diseases.

Healthy body cells divide naturally and eventually die off so that new, healthier versions of themselves can enter the body. Stem cells need to be healthy to make sure those copies are perfect; this maintains your body's quota of young, functioning cells. But where do senescent cells come into the picture? When your cells get to the point of toxicity, they're so dysfunctional that they don't go through the natural division and die-off cycles. Instead, they just float around like ghosts that recruit your other cells to join their toxic party. You can see why it's so important to keep both your stem cells and body cells from becoming senescent. Don't worry, we'll show you how.

THREE CHEERS FOR PROACTIVITY

All this talk about unhealthy cells might seem dire, but here's the good news: there are ways to keep toxic cells from stealing away our youth. The better our bodies are at fighting free radicals and repairing DNA damage, the better we are at staying young, healthy, and disease free. Our body has built-in systems to help us get the job done.

If there's one thing our body knows how to do, it's how to defend itself. Your immune system kicks in when diseases come knocking. Your skull, membranes, and fluids work together to protect that precious brain from injury. Likewise, your cells have a five-prong method that keeps premature aging at bay. But when your DNA damage is serious, our body's defense system can break down right along with it. The key to keeping your cells healthy is to strengthen the five anti-aging processes already at work inside them. Here's the quick rundown of your cells' five youthfulness boosters:

- **METHYLATION.** The cell's process of expressing genes; scrambled gene switching has been linked to everything from cardiovascular diseases to cancer.

- **INFLAMMATION.** The cell's defense against injury, usually in the form of swelling and pain; too much inflammation can trigger premature brain aging, Alzheimer's disease, arthritis, and memory loss.
- **GLYCATION.** The cell's way of regulating how blood sugar affects proteins and body fat; when faulty genes disrupt this process, you could develop a condition such as diabetes.
- **ANTIOXIDATION.** The cell's guard against those harmful free radicals; when too much oxidation occurs, the DNA damage could lead to tumors, cancer, and other serious signs of aging.
- **DNA REPAIR.** The cell's way of fixing damaged DNA; widespread DNA damage plays a role in cancer, a weaker immune system, and premature aging.

The Cell Solution is an all-natural lifestyle strategy that will boost your cells' defenses so you experience your healthiest years yet. It may sound simple, and that's because it is. If you know how to keep one cell healthy, you know how to keep all 100 trillion body cells healthy—no surgeries or unnatural chemicals required. All you need are the right diet, exercise, antistress strategies, and cell healthy nutrient habits.

If you know how to keep one cell healthy, you know how to keep all 100 trillion body cells healthy.

After decades of being told you're stuck with your genes, it's time to get in the driver's seat. With simple lifestyle changes, you can get those genes working in your favor. Your youthful today is only a few habits away.

THE TAKEAWAYS

• If your cells are healthy, your whole body is healthy.

• Healthy cells extend your window of youthfulness and slow the development of age-related diseases.

• When free radicals damage your DNA, your cells break down. The result? Premature aging.

• Keep your adult stem cells healthy so they can generate new, youthful cells in your body.

• You can control your genes and reverse aging by boosting your cells' natural youthful defenses.

COMING UP

• Learn the secrets to the cell-healthy diet.

• Discover why your cardio routine could be harming your cells.

• Tap into the ultimate stress-relief strategy.

EXPLORE MORE

- Visit www.youngertodaybook. com for animations and videos on cell aging.

2

The Lifestyle Solution

Life is busy. But we don't have to tell you that, do we? We're all in this hectic, twenty-first-century boat together. We wake up early. We grab a protein bar and a coffee on our way out the door—if we even eat breakfast at all. At work, we're stressed, frantic, and trying to juggle a million tasks and deadlines. By the time lunch rolls around, it's already 2:00P.M. Our smartphones are flooded with text messages from our family, emails from our boss, and Facebook updates from people we barely know. Between all this, we have to cram in a real meal and some exercise. We get to sleep too late, wake up too early, and start the cycle all over again.

Is it any wonder we're feeling fatigued?

We all know that lousy health can hurt your lifestyle. It's not easy to have a social life when you're constantly feeling tired! And don't even get us started on how your confidence suffers when you know you're aging and the world can see it. But what if we told you your lifestyle could actually be triggering those health issues? Your diet, exercise, nutrients, and responses to stress can either age you at lightning speed or keep you looking and feeling like the life of the party. We know which we prefer!

The first step to boosting your health is to get your diet, exercise, and stress-coping skills in tune with your cells' needs. Here's your youthful starting point.

GENES AND YOUR WORLD

The Complete Picture

Our genes, and the environment in which they exist, dictate our health and longevity.

Positive Environment		Improved Aging
+	=	+
Optimal Gene Function		Optimal Health and Quality of Life

Figure 2.1. Our environment plays a major role in our cell health. Luckily, this one is largely within our control.

Through the course of a typical, hectic day, your body collides with your environment in countless ways. There are the living things, like plants, animals, and the substances in our food such as bacteria. There are also nonliving things, such as the air we breathe, water we drink, and sun we absorb into our skin. The odd man out is stress, which can be caused by both biological factors and nonliving things like those harmful sun rays.

The first step to a youthful today is developing a lifestyle that minimizes the harm from all the factors we just listed. When we change our habits, we can change our genes. When we change our genes, we can recapture our youth. Your youthful lifestyle starts today.

OUR FOOD LIFESTYLE

When the going gets tough, our diet is usually the first thing to suffer. We all have our vices, but as a rule, Americans have many unhealthy ones. Our diet is sugary, processed, and a little too bountiful. We go for sweeteners—lots of them. We eat big meals. We grab the quick, processed food for a fast fix. We drink tons of coffee and soda. We pile salt on our mountain of French fries, America's veggie of choice. And soon, we feel our health take the hit.

We know what it's like when your body doesn't get what it needs. We've felt the stomachaches, the lack of energy, and the constant need to reach for pain relievers—not exactly the health routine we should be striving for. We've seen the wrinkles and the saggy skin, too. When you don't listen to your body, your food relationship gets one-sided pretty fast. Soon, your broken diet floods your system with free radicals that trigger the premature aging we were just telling you about. Like everything else, this big issue comes back to your tiny cells.

When you minimize free radicals in your diet, your cells stay stronger. When your cells are healthy, you're healthy. With the right strategies, your diet can serve up a big helping of youthfulness.

What We're Eating

How many times have you heard someone complain that they'd be able to eat all the junk food in the world if they had better genes? The bad news is that junk food is lousy for you no matter who you are. The good news is that our nutrition can actually cause our genes to help rather than hurt us. When we eat food rich in nutrients, such as free-radical fighting antioxidants, our plates can be a huge source of anti-aging health. Where in the grocery store have these cell-healthy foods been hiding? A lot of them are right in front of you. All you have to do is choose them over the processed stuff.

The typical American eats seventy-seven pounds of added sugar each year; that's a whole lot of glucose paired with a whole lot of refined carbohydrates. Glucose carries out many vital functions in our body, and without, it we'd be in trouble. And what would life be without a piece of cake every now and then? But let's not sugarcoat it: sugar is the villain of the story, and it all comes back to the type of sugar you're eating. When we eat and sip large quantities of sugar, the benefits go right out the window. The sugar that enters our body mingles with free radicals, and the result of this rendezvous isn't pretty. Those free radicals turn the glucose from sweet to toxic, and that sugar begins coating the surface of your cells. You can probably guess what happens next: these cells, now smothered in glucose, can't do their jobs properly. They age faster than normal and pump more toxic sludge into your body along the way. Add refined carbohydrates, like doughnuts, to the sugary mix, and the glucose gets into your bloodstream even more quickly.

When glucose throws off the harmony of your cellular symphony, your body will find plenty of ways to let you know. The first things you'll notice are blemishes or wrinkles—and no one wants those. But the visible signs of wear and tear are nothing compared to what's going on below the surface. When brain cells are sugarcoated, it can cause memory loss. When your immune cells are covered, that whole system is damaged, making you less able to fight diseases. This overly sugary state can also lead to diabetes, which causes health problems from head to toe. All in all, excess sugar jumpstarts all sorts of aging problems that you just don't want to deal with. Talk about sickeningly sweet!

And then there's the chemical issue. Processed foods have to be "processed" by something, don't they? That "something" is usually an unnatural chemical that adds even more toxicity to our bodies. These chemicals increase the amount of free radicals bombarding your cells, which forces your cells to work harder to stay healthy. This also gums up your DNA, which results in poor-quality cell copies. Since most of our processed foods contain white flour and sugar, you already know that your cells aren't in tip-top shape going into battle. What do tons of free radicals and weakened cells add up to? Messed up gene switching. This miscommunication throws off the whole symphony. Say hello to wrinkles, disease, and aging galore.

How Much We're Eating

The average American eats more calories than needed every single day—and at nearly 30 percent, our obesity rate is skyrocketing right along with our food intake. If the present predicament isn't scary enough, let's look at obesity down the road. Nearly 90 percent of the aging conditions we desperately want to avoid stem from being overweight. Hypertension, insulin resistance, cardiovascular issues, you name it: there are a million reasons why excess weight is bad for your health, bad for your cells, and bad for your youthfulness.

With all the starvation diets hitting the market these days, calorie restriction gets a bad rap. But there's a big difference between starving yourself and cutting down on your calories. Cutting down on your calories isn't about punishing or denying your body; it's about boosting your health. When done the right way, you can reduce your calories by 10 to 30 percent while still giving your body all the nutrients it needs; you've just got to choose healthy, unprocessed foods that naturally carry fewer calories. Smart caloric restriction can give you all the nutrients you need—plus the figure of your dreams and more healthy years to enjoy it.

The Food Solution

Our food supply isn't what nature intended, and it's not just the sugar. We're eating vegetables flown in from overseas, drinking water full of chemicals, and spending most of our grocery store budget on processed foods found in those dangerous middle aisles. We know how bad unnatural food makes your body feel, because we've been there. We also know that the sugary, processed stuff leaps out at you in grocery stores, shopping malls, movie theaters, and everywhere else you go. They don't call them "convenience foods" for nothing. Here's something to think about, though: you can go for convenience foods now, but how quick and easy is life going to feel when you're ailing in your fifties and sixties? We all have the power to be proactive with our dinner plates.

Think back to the last time you ate an apple. Did you stop to taste the sweetness? Did you notice the little burst of energy that hit your veins after the first bite? Did you feel the same lasting energy when you ate that junk food? We've become less aware, and our bodies have paid the price.

When we live off processed foods, we miss out on all the energizing nutrients that nature packs into each fruit and veggie. Our cells are left wanting more. The best way to give your cells what they need is to practice a diet based on whole foods. No, we're not talking about the grocery-store chain; we're talking about filling your plate with clean, unprocessed foods. When you process or cook your produce, the number of free radicals goes up. Meanwhile, many of the things we love about those ingredients are destroyed—including the enzymes and nutrients that made you grab that broccoli in the first place.

Thankfully, our diet is one thing we can control. When you eat food closer to its source, you won't just see the difference in your skin and your figure—though those are pretty nice perks. You'll also feel the change everywhere. Those nutrients will energize each cell from the inside out, and that healthy change will trickle down to everything you do. When you pay attention to how your food makes your body feel, the switch from processed to pure will be a no-brainer. It's transformed us, and it'll transform you, too.

We can also control whether we succumb to obesity and those other weight-related aging issues. Many times, it comes back to eating different foods. We're not recommending you limit yourself to carrot sticks; go ahead and breathe that sigh of relief. But as we said before, there are many benefits to cutting your calories by about 10 to 30 percent. Stock up on produce, lean proteins, and unprocessed foods that pack fewer calories but a bigger nutrient punch. When your cells are well nourished and not overfed, your whole body benefits—right down to the genes so many of us blame for weight gain.

How does cutting calories keep you looking and feeling youthful? It all comes back to your genes. When you trim down your portion sizes, it creates stress in your metabolism. Believe it or not, there's a good kind of stress, and this is it. The stress of eating less food triggers a group of genes that improves your body's responses to insulin and inflammation. The benefits don't stop there. With time, cutting back your calories by 10 to 30 percent can reverse insulin resistance altogether and make your cells much better at tackling stress. What does that all add up to? Longevity, youthfulness, and more years free from aging conditions.

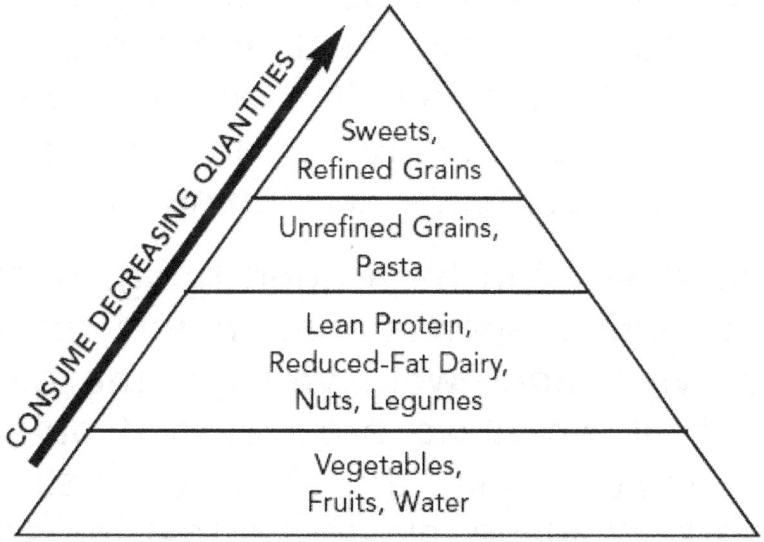

Figure 2.2. By filling your plate with lean proteins, fruits, and veggies, you can cut calories and boost your cell nutrition.

Cutting your calories every day is tough to do. We know how tempting those processed foods are! The key is finding that healthy middle ground, and pairing changes to your diet with the ideal combination of phytonutrients, natural chemicals found in plant foods that we'll tell you all about in the next few chapters. For now, here's your challenge: eat closer to the source. Fill your grocery cart with brown carbs instead of white ones. Choose seasonal fruits and veggies with the most intense natural color. Ditch the processed stuff as much as possible. Cut down on your portion sizes where you can. And most importantly, soak up every ounce of your new energy and vigor.

YOUR STRESSFUL LIFESTYLE

When stress hits, there's no mistaking it. Physically, you feel the pain in your chest and the quickening of your heartbeat. Emotionally, you're overwhelmed, irritable, and not exactly a joy to be around. Biologically, your cells are screaming out for help.

Sometimes, stress can be a good thing. It can trigger the fight-or-flight response that means the difference between survival and—well, you get the picture. The problem is that when we are stressed out over little things that don't actually threaten our livelihood (such as that work deadline or angry text message), the trigger starts hurting rather than helping us. One of the main villains in the stress story is cortisol.

When adrenaline hits, it activates the fight-or-flight response. That triggers the release of the hormone cortisol, which tells our cells to prepare for battle. But like everything else, cortisol is only good in moderation. When constant stress elevates your cortisol levels for too long, your body is thrown off balance, and the cells in your brain, immune system, and nervous system become scrambled. Elevated cortisol levels also deplete the cortisol supply that was supposed to be your saving grace in the event that a real threat arises. Oh, and did we mention that these cortisol surges can cause nearly every sign of aging in the book? Here's a quick list:

- Breakdown of the collagen and elastic tissues that keep your skin, joints, and muscles youthful
- Wrinkles, acne, and other unsightly skin issues
- Damage to your nervous system
- Increased risk for infectious diseases due to a weakened immune system
- Memory loss and decreased cognitive function
- Fat metabolism disorders and sugar cravings that threaten your youthful physique

With byproducts like those, it's no surprise "the stress hormone" also goes by another less-than-savory name: "the age-accelerating hormone."

While your body is creating cortisol and working extra hard to ensure your survival, it has a lot less energy for multitasking. You already know that your cells are dueling with free radicals around the clock to keep these volatile toxins from stealing away your youth. When you're stressed, your body is flooded with even more free radicals than usual. Your cells don't have as much energy for fighting, and the unfortunate result is the DNA damage that turns your cells from happy and healthy to damaged and toxic. In other words, a little more of your youthfulness slips away.

The Stress Solution

Stress ages you, plain and simple. But don't let that thought cause you too much anxiety. With the right coping techniques, you can fight off stress and get your genes working in your favor. Remember that fight-or-flight response we were just talking about? It applies here, too. Whether you're stressed from running around or from your looming work deadline, you have to activate the relaxation response—the opposite of fight or flight—to get back to a happier place. One big way to tap into this response is through meditation.

We all know that meditation techniques can ease your mind and foster that inner peace we're all searching for, but it turns out that this age-old practice can also work wonders for you biologically. So can yoga, its more physically active counterpart. Here's how it works: when you meditate or practice yoga, it triggers your relaxation response. That mechanism is your body's way of telling certain genes to switch "on" so that they can help transition your system from stressed out to harmonious. As the relaxation response sets in, many of the negative genes that switch "on" with stress are turned off, which lifts numerous roadblocks that were keeping your cells from functioning properly. Your cells can begin traveling freely along metabolic and oxidative stress pathways, which lets them get back to defending you against aging. This reopened highway system also gives your cells access to more energy for fighting free radicals.

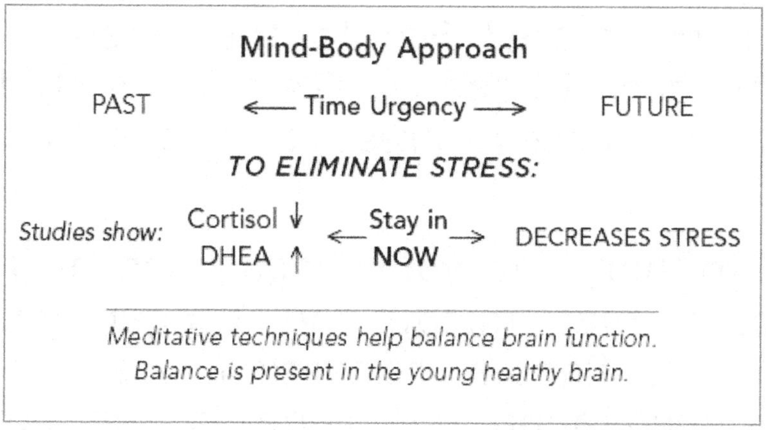

Figure 2.3. Meditation can stop the bad byproducts of stress, such as decreased DHEA levels. DHEA is an anti-aging hormone that's vital to your health. When you're stressed, DHEA production dwindles.

How does all this tie back to that youthful glow and healthy life? Stress produces cortisol and tons of free radicals. Both of these throw off your body's homeostasis, or balance. When that harmony is off-kilter, your cells scramble to get back to a happy place where they can function properly. Meanwhile, they're also fighting more free radicals than ever, which is even tougher than it sounds because they're already weak. Free radicals damage DNA, which can cause your cells to degrade and age you. Consider meditation your saving grace. Through meditation and yoga, you switch your body into relaxation mode. Your DNA will turn on the genes that get your body and cells back to good health, and you will ward off that sagging skin, weakened immune system, and weight gain. Your cortisol levels will go down, and the anti-aging stress hormone DHEA (dehydroepiandrosterone) will arrive to save the day. Want to know more? Flip to the references in the back of this book to see Dr. G's original study on how meditation decreases stress and boosts anti-aging hormones.

It's time to pull out your yoga mat or meditation pillow, focus that mind, and trigger your body's ultimate stress fighter: relaxation. We've included a beginner's meditation technique in the appendix of this book to help you get started on your mind-body journey. Go on, breathe that big sigh of relief.

YOUR EXERCISE LIFESTYLE

We all know the health benefits of exercise, but that knowledge doesn't seem to be getting us moving. Neither are our pricey gym memberships. But what if we told you exercise can change your genes and give you healthier, high-energy years? With the right kind of physical activity, your "golden years" will be golden again in no time. How's that for an incentive to get moving?

Walk into any gym in the country, and you'll most likely see a room packed with cardio machines, not enthusiastic gym-goers. Why is it that we never actually make it to the health club? Maybe it's because those cardio machines are prompting us to move in ways that just aren't natural. In nature, our ancestors moved either to hunt and gather or to survive. The survival part is triggered by the flight-or-fight response, which gives us that quick burst of adrenaline. Then the opposite mechanism, the relaxation response, kicks in to help our bodies recover from the stress we've been putting ourselves through. It's similar to how kids play in a playground. They don't run for thirty minutes straight, do they? Instead, they run for a few minutes, stop to rest, and then keep going. The same goes for the lion that stops to recoup, even as it's chasing down its four-legged main course. Nowhere in nature do we run hard for thirty minutes or more without pausing to rest. If we don't move like that in nature, we shouldn't exercise like that.

For the exerciser used to spending hours working out (and counting calories) at the gym, a shorter workout may sound like a cop-out. But when you exercise for more than forty minutes continuously, it actually hurts you more than it helps you. By stressing out your body, you're triggering cortisol, causing inflammation, and flooding your system with free radicals—the unhappy consequences of your prolonged stress. Meanwhile, your body has no rest period to help it recoup. What's a stressed-out cell to do?

The Exercise Solution: HIIT It

When it comes to exercise, our cells crave quality over quantity. Meet high intensity interval training (HIIT), also known as "cycles or bursts." It's your new personal fitness pal.

Never heard of HIIT? Here's how it works: rather than do that thirty minutes of cardio, you perform aerobic exercise at a very high intensity for thirty seconds to a few minutes, and then give your body a one- to five-minute recovery period of low- to no-intensity exercise. You repeat this a few times, and then you get to go home, have a healthful dinner, and get some sound sleep—or at least that's how it goes in an ideal world.

Let's zoom in and look at exercise from the cellular level. When we get moving, our blood hits our arteries and triggers a little spurt of nitric oxide, which relaxes your blood vessels. Then stress-fighting genes kick in to help your cells recover and fight free radicals. That's a good reaction. But a funny thing happens when you push your body to the brink with endurance exercise: If you exercise continually for an extended period, your body generates nitric oxide in a different pathway. This causes the opposite effect on your blood vessels, which tense up and stiffen. Then come inflammation and a continual onslaught of stress and toxicity. Your body is trying to tell you it needs to relax, but you're not listening.

High intensity interval training gives you all the strength, weight-loss, and heart-health benefits of endurance exercise, and like other forms of aerobic exercise, also improves your circulation and reduces your insulin levels. But the added bonus with the interval approach is that, unlike endurance exercise, the stop-and-go trains your relaxation response. The stronger your relaxation response, the better you'll be able to recover when a real fight-or-flight situation occurs. Think of those pauses between cardio as taking a moment to give your cells what they need. Your whole body will thank you.

How does HIIT tie back into your youthfulness? It's simple. By giving your body those rest periods, you're giving your cells the downtime they need to be healthy. With this relaxation response minimizing your stress levels, your cells are able to operate more efficiently. They're more adept at making sure your most beneficial genes get turned on, and we already know how that can thwart premature aging. Those other perks of HIIT—fat loss, cardiovascular health, and beyond—are just icing on the cake.

Your resistance training should follow the same "quality over quantity" regimen. Rather than perform more sets with fewer repetitions in each, push yourself to do more reps in each set, even if that means fewer sets overall. The higher intensity will increase your muscle mass more than the lower intensity routine, and it will also boost your bones' ability to take in key minerals. Just remember to give your muscles a few minutes of break between each set so that the relaxation can kick in.

Resistance training should be an important part of your new healthy exercise routine. It's been shown to improve glucose tolerance, and it also helps keep your insulin tolerance under control. But that's just the beginning of it. Weight training also can improve osteoporosis, strengthen your balance, and ensure your ankles, feet, and knees stay in top shape.

With the right combination of exercises, you'll look and feel healthier than ever before. For cell-health workout tips, be sure to check out the appendix in the back of the book. And do us a favor: consult with your doctor before you try any new exercise routine.

Your future is looking more mobile already.

THE NEXT STEP

When we first started our anti-aging routines, we were amazed at how quickly we saw and felt the changes. The skin tightened around that dreaded middle section. We finally had the energy to get through our days, which also made us a lot more fun to be around. Everywhere we went, people were leaning in and whispering, "Did you get a facelift?" They were shocked when we told them the answer was no. If you think you're feeling good now, just wait until you start getting those reactions from people!

We've given you the strategies that lead to a healthier you. Now it's time to get these tips in rhythm with your body. Let's stir up a little gene envy.

THE TAKEAWAYS

- Diet, exercise, and meditation habits can boost your gene health and keep you youthful.
- Choose unprocessed foods to give your cells the biggest boost.
- When you exercise, give your cells some relaxation time.
- Don't get stressed; it kills your mood and hurts your cells.
- Use meditation to reverse the negative effects of stress.

COMING UP

- Discover the importance of your body's biorhythm.
- Learn how to get your lifestyle and body in synch.
- Find out the healthiest times to eat, sleep, and exercise.

EXPLORE MORE

- Visit www.youngertodaybook.com for cell-healthy recipes, fitness routines, and meditation tips.

3

The Lifestyle Rhythm

You know how those stressed-out days, cardio binges, and infrequent, unhealthy meals affect your genes. You know why you need to keep your cells working to achieve that fountain of youth. Now it's time to bring your cells and lifestyle in synch. Get ready to tap into a healthy groove.

RHYTHM FOR ONE, RHYTHM FOR ALL

That hour or two on the couch can do wonders when you're fatigued, and eight hours of sleep can mean the difference between breakdown and bliss. But even when you're not moving, everything in your body is still working. Healthy, functioning cells are always scurrying around to keep the air moving to your lungs and the blood pumping throughout your body. They're constantly activating the genes that allow your body's mechanisms to kick in. While you're asleep, these worker bees are making sure your organs and tissues are getting the job done. They're also repairing their DNA.

With all this hustle and bustle going on, your body could become a mess pretty quickly without some biological direction. Luckily, there's something coordinating all your systems so the cells in your big toe work in harmony with the cells in your heart, fingertips, and beyond. That coordinator is your body's biorhythm, the master timekeeper that sets the pace for all the action in your body.

Here's how your biorhythm works: Your body has a master rhythm that presides over every cell, tissue, and organ in your body. The brains of that operation is a group of neurons that looks at whether it's light or dark outside, and then sets your body's clock to match where you are in the twenty-four-hour cycle. Once the neurons set this rhythm, they monitor all the tissues in the body to make sure each sticks to the schedule for sleep, feeding, and beyond. Believe it or not, there's also a clock in each of your 100 trillion cells. How do these trillions of clocks stay in synch? You have your tissues and organs to thank for that. Each tissue and organ coordinates the rhythms of its cells so that the specialized tasks we need to function are completed on deadline. When these secondary rhythms set by your tissues and organs are in time with your body's big clock, you're a healthy, well-oiled machine. When the rhythm breaks down, you begin to experience premature aging.

Think of your body's main rhythm as the conductor of the symphony orchestra we talked about earlier. The conductor knows the tubas, snare drums, clarinets, and violins all play different parts. His job is to make sure all these parts come together to create a perfectly executed song for the adoring crowd. If the players are having good nights, the lines and measures are executed without a hitch. But if that clarinet player is distracted and starts hitting wrong notes left and right, it's up to the conductor to try to get him back on track. A strong conductor will return the orchestra to harmony. If the conductor breaks down, so does the entire group.

Why does our biorhythm matter so much? That internal clock is our body's way of adapting to changes in our environment. When your body settles into a good rhythm with your lifestyle, the world is your oyster. You wake up every morning fully rested, get through your day with energy intact, and have time for quality meals and exercise. You're calm and at ease, and aren't plagued with any major health issues. But more often than not, our lifestyles don't take our body's advice. We go to sleep too late, eat huge meals when we should be resting, and fit in exercise just as our bodies are trying to wind down for the night. Then we expect our cells to deal with it.

A smooth biorhythm makes your cells happy campers. Your main clock knows when to "turn on" genes that help your cells function, and in turn, your healthy cells can keep your biorhythms in synch. But when your clock is thrown off, it can take a major toll on your trillions of tiny workers. With no clear clock to guide them, cells lose track of their workflow and scramble around trying to make up for that lost time. If those cells are damaged, they can throw off the groove of the other cells in their tissue. Soon, an entire organ or system loses time with your body's main biorhythm. Here's the takeaway: our cells rely on a strong biorhythm to stay healthy, and our biorhythm relies on healthy cells to stay in synch. We have to nurture this relationship. Don't worry, we're going to show you exactly how later in this chapter.

Unhealthy cells signal that more biorhythm problems are on the way. And because your main rhythm uses your sleep-wake cycle to set the pace for your body's activity, many health issues arise from not sleeping when your body wants you to. Over time, a chaotic sleep-wake cycle could lead to Alzheimer's disease. Dementia could set in if your body's clocks are out of rhythm across multiple systems for an extended period of time. Traveling frequently through different time zones can cause your cells to lose rhythm, and the cell damage and poor gene switching that ensues could increases your risk for cancer and other diseases. Over the long term, conditions like metabolic syndrome (a group of factors, including high blood pressure and high cholesterol, that boost your risk of heart disease and diabetes) and obesity can become your reality if your eating routine doesn't follow your body's sleep-wake cycle.

Still not relating? Think back to the last time you pulled an all-nighter. We all know how it feels to reach the point where staying out until all hours is no longer a great idea. When we're younger, our stem cells are willing to pick up the pieces and get our body's sleep cycle back on track. But as we age and our cells deteriorate, it can take days for our bodies to bounce back from one sleepless night. Whether we're talking about chronic diseases or that jet-lagged feeling, your biorhythm plays a huge role in your health.

The key to staying youthful is giving your cells what they need, when they need it.

The key to staying youthful is giving your cells what they need, when they need it. A functioning body clock makes for a happy, healthy you. It's time to get your diet, meditation and exercise habits in synch so that you can achieve your most youthful rhythm yet. Let's start with food, one of the lifestyle factors that can make or break your biorhythm.

YOUR FOOD GROOVE

In our body's perfect world, the rhythm that keeps us going revolves around the sun—literally. We'd be active during the day when the sun is out and recharge our batteries at night when the sun goes down. Aside from sleep, there's one other big factor that can trigger our cycle to shift: food. Our eating routine, or lack thereof, can cause our tissues to disobey that main rhythm and go their own course. Needless to say, this mutiny isn't so good for the health of the crew. For a prime example of how food can throw off your body's rhythm, just ask city slickers and shift workers.

New Yorkers and other city dwellers are infamous for their late-late night dinners. They get up early, work out, brave the commute, and grab a quick lunch whenever they can squeeze it in. Then it's more work. They finally get to dinner around 8:00P.M. or 9:00P.M., but by then, their body is already preparing for rest mode. Every once in a while, a late dinner is fine. We know you have a social life to maintain! But here's the thing: When you constantly eat when your body wants to be winding down rather than powering up, it's bad for your cells. Over time, your body will entrain itself to this unhealthy habit. Think of entraining as your body's way of adjusting its rhythm to a long-term change in your environment. Sometimes, this adaptation can help you survive a catastrophic event around you. But in the case of the late-night meals, this repositioning triggers some major hormone shifts.

Every once in a while, a late dinner is fine. But when you constantly eat when your body wants to be winding down rather than powering up, it's bad for your cells.

When we're in synch with our bodies, the appetite-suppressing hormone leptin peaks at around 2:00A.M. It stops food cravings throughout the night so that by the time we wake up, we're hungry and ready to give our bodies the energy they need to kick-start the day. But when you don't eat dinner until 8:00P.M., this appetite killer doesn't hit its high until

4:00A.M. You wake up in the morning and aren't hungry, even though your body could really use the energy. Every part of your body is connected, so this shift in your eating pattern quickly throws off your energy, metabolism, and beyond. It also messes with your ability to get a good night's sleep, which is exactly what your body needs to keep your internal clocks moving in rhythm. Talk about falling out of step.

Shift workers are another case of biorhythm gone awry. Our body wants us asleep when it's dark outside and active when the sun is up. In shift workers, who stay active late into the night, the food-sleep-light three-way gets misaligned. It's the same story as with the city dwellers. Shift workers tend to eat less during the day and have their biggest meal late at night—and can we really blame them, given the hours they're keeping? The problem is, the body isn't designed to receive the biggest boost in energy at an hour that should be sleep time. Because of this eating change, shift workers' sleep, energy, and metabolism patterns also move away from the norm. The unfortunate result is that shift workers are much more likely to develop metabolic syndrome than the typical nine-to-fiver is. They're also at higher risk for obesity, the official gateway to 90 percent of aging conditions. And because they are staying physically active so far into the night, their bodies have a tough time adjusting to the "sleeping during the day" routine. From there, the dominos keep on falling.

What do the city dweller and the shift worker have in common? They're both at risk for a metabolism meltdown. Every day, your body's genes play a hand in regulating the activities of your metabolism. Gene receptors make sure your metabolism and body's clock coordinate with each other, but this communication doesn't do much if you get into the habit of eating during your natural rest periods. Your genes aren't sure what's going on, so their tissues shift rhythm to match your abnormal eating habits. This throws your systems—and metabolism—out of synch with your big clock. There's a reason why insomnia and the eating disorder night eating syndrome (NES) are such concerns. When your nocturnal eating habits prevent your body's systems from working in harmony, it can kick-start aging issues like obesity, diabetes, and a wide range of metabolic disorders. That's one quick ticket to body breakdown.

This leaves us with a burning question: when should we eat? Everyone is different, but what we know for sure is that we're a lot better off when we're working with our body's rhythm. Be kind to your body by not eating at hours when you should be sleeping. Eat your biggest meal in the mornings and your smallest one at night to prevent metabolic disorders. Eat most of your carbs in the morning, rather than at night, so your body has a chance to use the energy they provide. Eat smaller meals more frequently to strengthen your energy balance so that you have the vigor to get through your day. But above all else, establish a sensible eating schedule—frequent meals during the day and no 2:00A.M. binges—to help your body set a consistent pace.

In the previous chapter, we told you all about the ways in which our modern diets are harming our cell health and hitting us with major free-radical damage. You should be staying away from loads of sugar and processed carbs, and should opt instead for foods that are in their natural form. But that's only half the equation. When you eat carries just as much weight as what you eat. By applying our cell-healthy diet tips in a sensible eating schedule, your body will reap the most health benefits. The sooner we get our food and biorhythm dancing in synch, the sooner our body and diet will stop stepping on each other's toes.

YOUR STRESS GROOVE

Prolonged stress isn't ideal any time of day, whether you're exercising at the gym or fretting over your latest family tiff. But the good news is there's a way to turn around your stress and strengthen your biorhythm at the same time. It all comes back to meditation's cell-health powers.

We know what happens when you are stressed: the cortisol flows like wine. As this aging hormone floods in, it throws your system off balance and scrambles the activity of your cells. Over time, these elevated cortisol levels lead to almost every sign of aging in the book. Your skin suffers, your memory gets fuzzy, and your immune system has to work way too hard to ward off diseases. Your biorhythm also shifts. Where there was once harmony, there's now a complete mess. Cortisol sure knows how to ruin a good thing.

Your biorhythm is your body's way of reacting to bigger environmental changes. Think of your stress system as your body's way of adapting to the smaller, unpredictable stuff. Just like everything else, it's governed by your biorhythm. Your cortisol has a natural cycle, ebbing and flowing at different times of day. When you're chronically stressed, cortisol surges, and your natural rhythm shifts from the happy norm. What comes next are sleep issues. The second wave of uninvited guests are obesity, metabolic syndrome, and major cardiovascular complications that steal away your youthful years.

But don't let that thought stress you out just yet. This is another case when meditation can save the day. Meditation has been shown to decrease cortisol and to synch your body's clocks. How does this mind-body strategy work its wonders? It starts with melatonin, a hormone whose official duty is to get your rhythms back on track. When you meditate, it boosts the amount of melatonin in your blood's plasma. This hormone immediately goes to work regulating your sleep-wake cycles. Your biorhythm gets moving in the right direction, and you get a better night's rest—and we know how important a solid sleep schedule is to your body's healthy groove.

Melatonin isn't just a sleep regulator. It's also a protective older brother to your cells. When melatonin kicks in, it nudges a bunch of antioxidant genes, telling them to switch "on" so that they can help your cells fight free radicals. In this way, melatonin boosts cell survival and slows that slippery slope into cellular aging, all the while getting your cells' rhythms back on track. Consider your stress alleviated.

YOUR EXERCISE GROOVE

Just about every day, we hear about a new exercise program that promises to give us the most toned, conditioned bodies of our lives. You know the ones we're talking about. If you're like most Americans, you've probably tried one or two of them without getting any real results. It's time to stop listening to the fitness gurus and start listening to our bodies.

Our anti-aging fitness plan doesn't require gadgets or expensive equipment. It doesn't tell you to spend your life at the gym, either. All we're telling you to do is perform exercise in a way, and at a time, that has your cells' best interests in mind. Here's a bit more on how your biorhythm and fitness finesse align.

Your body goes through many cycles in its twenty-four-hour day, and many relate to how well you'll perform once you start moving. Your body temperature and arousal levels are both lowest in the morning as you're waking up. That means that for many people, the best time to perform skills-based exercise—anything requiring fine motor skills and balance—is right after you switch off that alarm and start your day. There's also a general connection between biorhythm and sports performance: just ask those pro athletes, who have been shown to perform a little less than brilliantly when they have to repeatedly play "away games" in different time zones. This may surprise you, but when you think about the way our biorhythm works, it makes perfect sense. Traveling through different time zones forces our bodies to adjust to new sleep schedules. But with athletes, the stay in new cities is usually brief at best, meaning that they never quite make the adjustment. Our bodies can't perform to their best ability if our cells are scrambling around to get us back into a happy rhythm, and athletics demands that many parts work together in a sophisticated way. This scenario gives a whole new meaning to "home field advantage."

Exercise with your body's rhythm rather than against it.

That's too bad for those professional athletes, but how do all these little pieces work back into your healthy exercise routine? It's simple: you need to exercise with your body's rhythm rather than against it. Exercise anytime from wakeup until early evening, but know that after lunch your body wants to take a natural relaxation period. This is yet another reason to love siestas! Take a quick walk in the mornings to get your blood flowing so that your body can energize itself for the day ahead. And please don't push your body to its highest intensity around 8:00P.M. or 9:00P.M., when it is trying to wind down and transition into sleep mode. In short, move when your body wants to be moving.

Another way of giving your cells what they need when they need it is to tap into your relaxation response. We've already raved to you about HIIT, but it's worth mentioning again. After a few minutes of exercise, your cells are begging you for a break so they can de-stress. Giving them that pause between intense movements is the ultimate way to nurture your biorhythm. When your exercise values the movement and the recoup periods equally, it helps all the parts of your body function more efficiently and in perfect time. That symphony has never sounded better.

YOUR YOUTHFUL GROOVE

When you start working with, not against, your body's rhythms, you'll be amazed at the difference it makes in your physical, mental, and emotional well-being. We know because we've been there. You'll notice the spring in your step and the boost in energy that makes each day feel like an adventure. Your mood will improve, your muscles will be tighter, and you'll be motivated to find more ways to share your new outlook with the world. And here's the best part: the more you listen to your body's rhythms, the easier it gets to give it what it needs when it needs it. It's the same for eating food, exercising, and de-stressing; a little bit of awareness really helps us understand and respond to our cells' needs in the right way, at the right time.

Ditching the processed food, practicing more sensible exercise, and meditating away stress will get you on the right track for rediscovering your youth. We know these things work because they've transformed us from the inside out. But we also know that sometimes, life gets in the way. We're human above all else, and when life gets messy, it can be tough to maintain our new proactive habits. So what's a youthfulness-seeker to do when things get hectic? Don't worry; we've got you covered.

We offer a recommended mix of nutrients that will give your body a healthy boost even when your lifestyle is lacking. These cell-boosting phytonutrients are natural, scientifically proven ways to slow down aging and reacquaint you with your more youthful years. They'll give you more nutrients than you could ever fit into your diet, and they'll get your biorhythm back on track when your sleep, diet, and exercise just won't align. And here's the best part: these all-natural complexes fit in right with your on-the-go lifestyle. Say hello to better sleep, more energetic days, and countless healthy years. The Cell Solution recommendations will get you there. Get ready for your youthful today.

THE TAKEAWAYS

- Get your lifestyle in synch with your body's biorhythm to ward off premature aging.
- Avoid eating late at night to keep your sleep-wake cycle and eating routine in harmony.
- Meditate to decrease stress and promote a healthy sleep pattern.
- Take your healthy lifestyle to the next level with our recommended phytonutrients.

COMING UP

- Get to know your cells' natural weapons against aging.
- Learn the secret to repairing your genes from the inside out.
- Find out how nutrients can make you feel younger today.

EXPLORE MORE

- Visit www.youngertodaybook.com for videos on biorhythm and beyond.

4

The Cell Solution Road Map

You have the exercise routine down, you know how to flex that meditation muscle, and your food relationship is looking more wholesome by the minute. What more could your body possibly need? Well, a lot. Your cells have very particular nutritional needs, and although eating from the source can get you moving in the right direction, the real anti-aging magic happens when you cross the finish line. That's where phytonutrients come in. With the right combo of vitamins, minerals, digestive enzymes, and phytonutrients (the nutrients found naturally in herbs, plants, and vegetables) you can give your cells all the nurturing they need and deserve. The best part is that these natural complexes won't just stop aging in its tracks; they'll also reverse a lot of the aging you've already experienced. When you combine your diet, exercise, and stress habits with these phytonutrients, you've embarked upon the ultimate cell-health routine.

But hold on a second. We know what you're thinking. What if you're in the prime of your youth? Even if anti-aging isn't on your radar yet, phytonutrients deserve a place in your supplement plan. The beauty of these all-natural nutrients is that, although they hold a plethora of anti-aging benefits, they also improve your health no matter how old you are. Healthy cells make you healthier from head to toe, and good health is something we can never have too much of. It's never too soon to start thinking about how you can prevent premature aging.

You've heard all about the lifestyle solutions, but now it's time to take it to the next level with phytonutrients. These groups of all-natural, science-based nutritional supplement recommendations will get your genes working in a more youthful way. Whether you're sixty and need a youthful boost or twenty and want a healthier lifestyle, these complexes are your new go-to health tools. Welcome to health 2.0.

BEYOND VITAMINS AND MINERALS

Vitamin B12. Zinc. Magnesium. Riboflavin. After a while, sorting through vitamins and minerals starts feeling a lot like alphabet soup. With all the supplements lining your pharmacy's nutrition aisle, how are you ever supposed to know what your cells actually need to stay healthy?

We have a lot of good news for you. First, you and every other person in the world is 99.9 percent the same when it comes to DNA. That means that there's a core group of nutrients that will help every person's cells function at the top of their game. If you know how to keep that cell in your lung healthy, you know how to keep its peers in the heart, brain, liver, and beyond doing their best work. The same goes for your stem cells. Now for the second piece of news: we've finally moved beyond all those individual vitamin tablets. Vitamins and minerals are great at keeping our cells functioning, but on their own, they don't do much to slow down our cells' toxic train wreck. If you're driving in your car and see your youthfulness in the rearview mirror, you wouldn't just hit the brakes, would you? What you should really be doing is driving in reverse. Phytonutrient complexes will get you moving backward—in a good way—by slowing aging at its roots.

Vitamins and minerals are great at keeping our cells functioning, but on their own, they don't do much to slow down our cells' toxic train wreck. Phytonutrients can slow aging at its roots.

By combining vitamins, minerals, digestive enzymes, and phytonutrients, we can take cell health to the next level. These complexes don't just give you what you need to get by; they actually delay your cells' transition from healthy to toxic by turning the right genes "on" and "off." The secret is in the combination of nutrients, and phytonutrients are the star of the show.

Phytonutrients are transforming the way we think about supplements and are debunking the myth that we can't control our genes. Kiss your once-a-day vitamin habit good-bye, and make these complexes part of your youthfulness routine.

PHYTONUTRIENTS: VITAMINS AND MINERALS 2.0

You've heard of superfoods, right? Think of phytonutrients as supernutrients. These natural nutrients are found in plants, herbs, and vegetables, but they pack a powerful health punch in humans as well. Filled to the brim with nutrients, these phytowonders have been known to fight diseases ranging from cancer to diabetes. Phytonutrients can also turn on or turn off certain genes, which makes them one of the most potent anti-aging tools around. Our favorite thing about them is that they work all this magic without putting a single unnatural chemical into your body. Talk about an impressive resume.

If phytonutrients still seem like a foreign concept to you, try thinking about them this way: taking phytonutrient complexes gives your body all the nutritional benefits of drinking gallons of cold-pressed fruit and veggie juice, and much more. All that's missing is the messy cleanup! With phytonutrients in your daily routine, you can ensure your body gets all the plant-based nutrients it needs to slow down your aging and kick-start your health from the inside out—even on those "off" days when your healthy diet goes out the window. You don't have to wait to see and feel the results, either. These youthfulness defenders leave you feeling energized almost immediately, with no unnatural chemicals or nasty side effects in sight. You'd have to eat a mountain of produce each day to get all these benefits from your diet alone.

The food you put into your body has a tremendous impact on your cell health. So do exercise and stress levels. But here's the thing: if adding phytonutrients to your daily routine can slow down your aging, why wouldn't you do it? We're not recommending some new prescription drug; we're giving you a proven, natural way to get your genes operating in your favor.

When we give our bodies the right combo of phytonutrients, we reap the benefits in six youth-boosting ways: our cells achieve a more balanced metabolism through caloric restriction, feel their biorhythms get in-synch as they become regulated, gain stronger DNA repair skills, become better equipped to fight free radicals, have healthier stem cells, and have telomeres less prone to premature shortening. Consider this your new youthfulness road map.

CALORIE RESTRICTION

We already threw the scary figure at you: 90 percent of all age-related health issues stem from being overweight. To mitigate obesity, diabetes, and major metabolism disorders associated with excess pounds, you need to get your cells' metabolic activity back on track. In this arena, calorie restriction is your new best friend.

Caloric restriction with optimal cell nutrition isn't a miracle worker, but it's pretty darn close. When you cut back on your calories by 10 to 30 percent for an extended period of time while consuming top-notch nutrients, it sets off a positive chain reaction that helps your body adapt to stress like a pro. Some of the biggest perks are better regulation of your blood sugar and a boost in your insulin sensitivity. Here's why that matters: when your sugar and insulin levels spike, it can damage your cells and, in the process, switch on harmful genes and turn off beneficial ones. We know that good gene switching can delay the development of aging diseases, so it shouldn't surprise you to hear that calorie restriction can slow the onset of cancer, diabetes, cardiovascular diseases, and other metabolic issues.

If all it takes is cutting back on your food, why do you need a nutritional supplement? It's not just about calorie restriction; it's about giving your cells optimal nutrition at the same time. Phytonutrient complexes will do that better than any daily diet. Also, cutting your calories by 10 to 30 percent every day for the rest of your life isn't exactly a piece of cake—and that's without trying to boost your nutrient levels while you do so. If we're not careful, consistently trimming calories can make us feel moody and deprived, which is the recipe for unhealthy bingeing. That's why we're such big fans of phytonutrient complexes. With the right mix of phytonutrients, vitamins, minerals, and digestive enzymes, you can get an effect similar to caloric restriction—without the mood swings and sugar binges!

CELL REGULATION

Your lifestyle can do a lot to get your body's symphony in synch, but phytonutrients take it to a whole new level. With phytonutrients on your side, the players in your orchestra become world-class musicians, and the notes come together in perfect harmony.

Your cells have a natural rhythm for performing the repairs that keep them healthy. And there's no surprise here: for you to be healthy, that rhythm has to stay in time with your main body's cycle. Cells generally perform their repairs at night while you're sleeping and go about the rest of their functions during the day while you're most active. The problem is, once you turn forty, your cells start getting a lot less nimble at their timekeeping. Why does this happen? As we age, the system that governs our cells' active periods, the sympathetic system, and the one that governs their passive periods, the parasympathetic system, are thrown out of alignment. Where that occurs, unhealthy cells are sure to follow. So is premature aging. What's an out-of-synch cell to do?

It's phytonutrients to the rescue. These mighty nutrients don't just fight disease or fine-tune gene expression, they also regulate the hormones that keep your cells' rhythms going strong. Your cells have different needs depending on the time of day. We know that when morning rolls around, our cells need a boost of energy. In the evening, it's all about helping them best repair their DNA and helping you get some sound sleep. The same internal clock governs our cells' hormone needs. Growth hormones, for instance, should kick in at night, while thyroid hormones are most useful to you during the day. By balancing out your hormone levels and strengthening your cells' internal clocks, phytonutrients can get your cells and the body's parasympathetic and sympathetic systems back in synch with your body's needs. That sounds like harmony to us.

Our maintenance recommendations don't just fix the cell's repair cycles, they also keep your cells healthy from their insides right on out to their membranes. Think of it as the best tuneup you've ever had.

DNA REPAIR

We don't mean to sound like a broken record, but DNA repair is one of the keys to keeping your cells healthy. Your body knows this, so it's supplied your cells with a built-in fix-it shop where DNA comes to be mended. But, as you can imagine, unhealthy cells aren't as good at fixing DNA as healthy ones, which means that more of the genetic matter that comes through their shop doors doesn't leave mended. The same goes for your stem cells, which can't make good copies to replace your dying cells if their DNA is all scrambled. Damaged DNA leads to less-than-perfect gene switching, and one key to aging gracefully is to make sure your helpful genes are active and the harmful ones are inactive. If these damaged cells created a "good health" wish list, they would undoubtedly make DNA repair tools their top priority.

If you don't have the proper nutrients in your diet, it can subject your cells to all kinds of wear and tear. But the good news is that phytonutrients can also reverse the damage by helping your cells become better DNA mechanics. Our recommended combos (outlined for you in Chapter 5) will turn your cells into top-notch genetic repair shops. Believe us, your cells will thank you.

SUCCESSFUL REPAIR	UNSUCCESSFUL REPAIR
Radiation, toxins, poor diet, and environment damage DNA strand.	Radiation, toxins, poor diet, and environment damage DNA strand.
Repair enzymes and protein process repair DNA. No damage evident.	Repair enzymes and protein process fails. Damage evident.
DNA copies itself making accurate cell copies.	DNA copies itself making poor cell copies.
Healthy Cell and No Disease = Optimal Aging in Your Cells	Unhealthy Cell, Chronic Disease = Accelerated Aging in Your Cells

Figure 4.1. Good DNA repair equals better cell copies, healthier cells, and a more youthful you.

FREE-RADICAL SCAVENGERS

Nothing steals away your youthfulness quite like free radicals. See those wrinkles on your face? You have these volatile little molecules—free radicals—to thank for them. Feel that saggy skin in your tummy? It's often the same culprit. When it comes to cell health, free radicals deserve a lot of attention.

When free radicals from the environment bombard your cells en masse, it can do a real number on your DNA. The story usually goes something like this: DNA is damaged and can't keep track of which genes should be switched "on" and which should be "off." Your bad genes start popping up in major ways—you know, that premature aging we all want to avoid. Meanwhile, your cells are weakened from the inside out and just aren't the skilled free-radical fighters they were in their prime. When the aging cells start turning toxic, they muck up the rest of the neighborhood as well. This situation calls for some serious damage control.

Our cells have a built-in system to help them battle toxins. Most of the fuel for that oxidation system—the toxin-fighting antioxidants—comes from the food we eat. But when free radicals bombard our aging, unhealthy cells, it's no longer a fair fight. You'd have to eat mountains of leafy greens and orange veggies to come out victorious.

Thankfully, there's a hero who can shift the odds back in our cells' favor. Meet free-radical scavengers, antioxidants that stop free radicals before they can turn your DNA, and cells, toxic. They do it by destroying the extra electron that makes free radicals so volatile. Free-radical scavengers are found in numerous phytonutrients, which can give you all the benefits of platefuls of green and orange veggies, and more. Because antioxidants work best in diverse groups, the natural supplement route is an easy way to make sure you get the ideal mix of these health-boosters.

We'll give you the full rundown on which phytonutrients fight free radicals in the next chapter, but until then, swallow this: with the right combination of antioxidants, you'll be able to boost your cells' aging defenses and keep your genes switching in your favor. That's two big thumbs up for your health.

STEM CELL MAINTENANCE

The more healthy stem cells you have, the more your fountain of youth is flowing. It's time to get proactive about keeping your stem cells healthy.

Aging cells can muck up the whole neighborhood, but when your stem cells become dysfunctional, the toxicity story is even more dangerous. As we explained before, your stem cells can essentially revive your body's cell population by making healthier, younger copies of your old, dying cells. Think of stem cells as crucial body backup. When your stem cells are healthy, the shiny, happy cells they create improve your body's health. But when your stem cells suffer DNA and free-radical damage, their copy making goes awry. Each cell copy resembles the original a little less, until it's hard to remember what functioning DNA even looks like.

From your thirtieth birthday onward, you'll always have some aging stem cells in your body's population. The big problems arise when your aging stem cells start turning their nearby buddies toxic. When this happens, your muscle mass will decline, and the fat, collagen, and elastin that keep your skin full and smooth will start to vanish. Your hormone levels will decrease. Then your kidney function weakens, and bacterial and viral infections will start to creep in through the cracks in your immune system. Welcome to premature aging.

Your stem cells age just like your body cells do, so the free-radical fighting and DNA repair are still a big part of the picture. But stem cells are one of the main defenders of your youth, so they deserve care that goes a little above and beyond. That's where phytonutrients come in. Certain plants have an uncanny knack for boosting your stem cells' proliferation, renewal, and overall activity levels. Our recommended cell health supplements give you the perfect combination of nutrients, paired with enzymes, that make sure your cells get every last drop of the benefits.

TELOMERE MAINTENANCE

If there's one thing we all dread, it's being asked how old we are. Like us, our cells have an ID that reveals their age to the other cells they meet. If you come across a cell with long telomeres (the end caps on your chromosomes), you know it's pretty young. If you meet a cell whose chromosome end caps are short, you know it's already been around the block a few times. When these little end caps get so short that a cell can't divide anymore, it sends a signal that tells the cell it's time to die. Nobel Prize–winning research has shown that keeping telomeres healthy could be the key to longevity and youthfulness. Why don't we just figure out how to keep our telomeres from shortening? We thought you'd never ask.

Our telomeres shorten each time our cells divide, but these little end caps also are truncated prematurely when we put our cells through lots of stress. Cortisol, that hormone that kicks in along with our fight-or-flight impulse, can cause those telomeres to shorten much faster than they should. So can inflammation and that dreaded DNA damage. Practicing meditation and eating fewer processed foods can cut back on DNA damage and free radicals, but we now have the science to take it a step further. The right combination of phytonutrients can trigger hormones that stop each of these telomere-shortening processes at the root. The result? Your cells stay younger for longer than ever before, and so do you.

PHYTONUTRIENTS IN ACTION

We've come a long way since vitamin tablets first came on the market. For the first time, we have the science to combine phytonutrients, vitamins, and minerals in a way that can actually trigger the genes that help your cells stay healthy. And the best part is that every single one of these phytonutrients is natural and free of side effects. We've shown you some of the reasons why your cells will love phytonutrients, but now it's time to experience one more: the way it feels when your health improves from the inside out. When you combine the recommendations in the next chapter with your exercise, diet, and stress habits, you'll tap into the ultimate level of cell health.

THE TAKEAWAYS

- We've moved beyond vitamins and minerals to provide our cells with complexes that can slow the spiral into aging, while boosting cell health like never before.
- Phytonutrients are natural disease fighters found in plants, herbs, and vegetables.
- By switching certain genes on and off, phytonutrients can slow down premature aging.
- Phytonutrients can boost your cell health in six ways: increasing DNA repair, fighting free radicals,

maintaining your stem cells, preventing your telomeres from shortening prematurely, balancing your metabolism, and getting your cells' biorhythms in synch.

- To get the best anti-aging boost, pair phytonutrient complexes with cell-healthy exercise, diet, and stress-coping habits.

COMING UP

- Discover the combination of phytonutrients, vitamins, and minerals that will make you look and feel younger today.

EXPLORE MORE

- Visit www.youngertodaybook.com for videos, illustrations, and other multimedia content on cell health.

5

The Cell Solution in Action

Toss out your once-a-day vitamins and say good-bye to those recommended daily intakes. There's a better, natural supplement approach on the block.

The Cell Solution is easy, all natural, and proven to send premature aging to the hills. Yes, there are many ingredients in these complexes, but don't let the long list and multisyllabic names fool you: to ensure your body is getting the best, cleanest boost, each phytonutrient is natural and cold-pressed (meaning it is extracted without being "cooked" so that more of its natural enzymes remain). Taking these phytonutrients is the equivalent of juicing mountains of fruits, veggies, and herbs. You couldn't eat this amount of cell-healthy nutrients if you tried! When you take these all-natural nutrients, you also know that your source of nutrition isn't tainted by chemicals in the food supply. And there are even more perks: these recommended complexes will give you the perfect serving of cell health, plus a digestive boost to make sure you soak up every drop of that goodness. What does all that add up to? A natural anti-aging supplement strategy that puts you in control of your aging. It's time to inspire a little gene envy.

Although these phytonutrient recommendations are indispensable anti-aging tools, these nutrients are beneficial to you no matter how many candles are on your birthday cake. Whether you're thirty and seeing your first wrinkles creep in or sixty with a lot of life left to live, this magic mix of phytonutrients, vitamins, minerals, and digestive enzymes will help you put your most youthful face forward. We saw and felt the difference overnight, and you will, too. When was the last time you said that about a multivitamin?

HOW IT WORKS: FORGET ABOUT ONCE-A-DAYS

Before we get you started on our recommended anti-aging supplement plan, let's cover all the basics. You already know our suggestions work because they combine all-natural phytonutrients, vitamins, and minerals. You also know that you should pair these nutrients with digestive enzymes that help you absorb every drop of those benefits. Now it's time to get down to the logistics of how and when to take these nutrients so that you get the biggest health boost.

Here's the nitty-gritty on these cell-healthy nutrients. We recommend that you take these six complexes together in the exact amounts we've listed. The mix of ingredients is what helps create the synergy your body needs to switch good genes "on" and bad ones "off." If you only take a few of these phytonutrients here and there, it will still benefit your body—after all, natural, herbal, and plant-based nutrients can never hurt. But if you only commit to part of the list, you're also letting a lot of the cell-health nutrients slip by the wayside. The same goes for the morning and evening recommendations. The morning combinations will jump-start your energy, and the evening ones will help your cells repair themselves as you drift off into the best sleep you've had in ages. We suggest taking both to maximize your benefits.

Taking these phytonutrients is the equivalent of juicing mountains of fruits, veggies, and herbs. You couldn't eat this amount of cell-healthy nutrients if you tried!

With any supplement plan, some people always think it's too much trouble to assemble all the required nutrients. We understand where you're coming from, and we know that you're busy. But we also know these recommendations are worth the effort. If taking daily supplements will help you look and feel better, isn't it worth a few minutes out of your day? Taking these phytonutrients is the equivalent of juicing mountains of produce and many less-than-common roots and herbs. When you compare it to the dietary alternative, this plan sounds easier already. Most importantly, these recommendations will give your cell health a bigger boost than diet alone ever could. It works, plain and simple.

Before we send you on your new cell-health adventure, let's talk a bit about those recommended daily intakes

(RDIs) you see on food nutrition labels. You may be wondering why our plan doesn't stick to those values, and here's why: RDIs were developed to give you a ballpark figure of what micronutrients the average healthy person needs in a given day. However, most people's diets aren't offering a bounty of nutrients, and those RDIs also don't take into account your genes. Even if RDIs did include what we need for cell health, most of us wouldn't be able to absorb all those nutrients anyhow. That's especially the case for people over sixty, the age when our absorption abilities really start to dwindle. This is one of the big reasons why we've outlined suggested digestive enzymes for you.

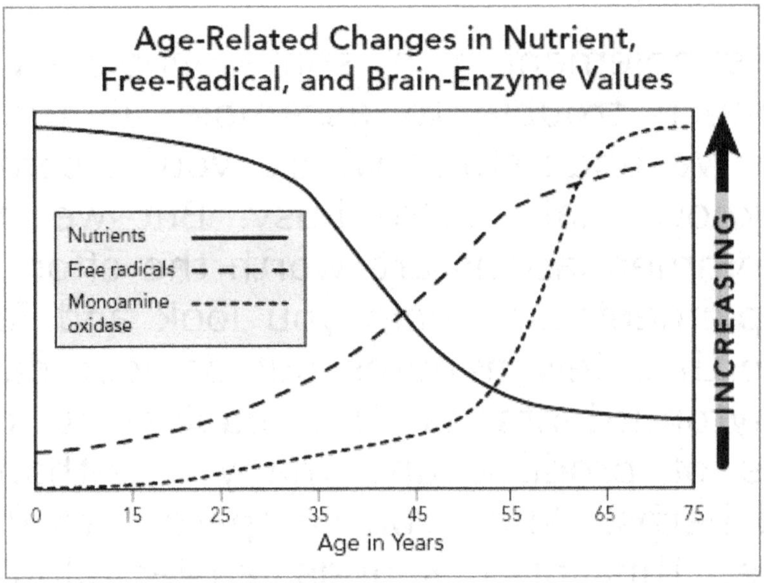

Figure 5.1. After age sixty, your abilty to absorb nutrients takes a nosedive.

We're taking supplements to the next level by recommending the daily nutrients you need to slow down aging and keep your cells healthy. When you pair that with the diet, exercise, and meditation techniques we've outlined, you'll reap the full benefits of the Cell Solution lifestyle.

We've moved beyond vitamins, and we've moved past RDIs. Care to join us? Here's all the information you need to kick-start your youth today.

CALORIE RESTRICTION COMPLEX

When you take this group of natural supplements, you'll get all the benefits of cutting calories and more. A powerful pair of carnitines, both found naturally in grains, will boost your blood flow as they work wonders on your gene switching. Coenzyme Q10, which may sound like a chemical but actually comes from botanicals, will fight free radicals and balance your energy production. Combine these phytonutrients with the other recommendations in the calorie restriction bundle, and your metabolism will be better than ever.

TABLE 5.1. CALORIE RESTRICTION COMPLEX	
MORNING CELL SUPPLEMENT RECOMMENDATIONS	
Trans-resveratrol	15mg
Pterostilbene	0.8mg
Fisetin	0.8mg
Alpha lipoic acid	20mg
Coenzyme Q10	2mg
Betaine HCl	1.6mg

TABLE 5.1. CALORIE RESTRICTION COMPLEX	
Sulfur from MSM	1.2mg
L-Carnitine tartrate	4mg
L-Carnitine HCl	2mg
EVENING CELL SUPPLEMENT RECOMMENDATIONS	
Trans-resveratrol	15mg
Pterostilbene	0.8mg
Fisetin	0.8mg
Alpha-lipoic acid	20mg
Coenzyme Q10	2mg
Betaine HCl	1.4mg
Sulfur from MSM	1.2mg
L-Carnitine tartrate	4mg
Acetyl-L-carnitine HCl	2 mg

CELL REGULATION COMPLEX

Think of the cell regulation complex as a much-needed housecleaning for your cells. If you've been meaning to work more teas into your life, you'll love these recommendations. Use cold-pressed extracts from those nutritional teas you love—white, oolong, black, and green—so that you get all their energy benefits without sipping endless cups. These colorful teas are a powerful, natural way to guard your body against the free radicals that throw it off-kilter. Another signature piece of the puzzle is a mushroom extract, which has been used in Chinese medicine for ages to help cells survive less than favorable conditions. And because we know that your stressful, fast-paced lifestyle often keeps you up at night, we recommend using the herbal sleep blend to promote restful, uninterrupted sleep.

Sure, you could find all these ingredients in food, but think about how much you'd have to eat and drink to get the same cell-health value that these extracts deliver.

TABLE 5.2. CELL REGULATION COMPLEX	
MORNING CELL SUPPLEMENT RECOMMENDATIONS	
Red algae extract	60mg
Gotu kola leaf	6.7mg
Green tea leaf extract, oolong tea leaf extract, white tea leaf extract, and black tea leaf extract	67mg
DMAE bitartrate	10mg
Cordyceps sinensis mushroom extract	5mg
Green coffee bean extract	50mg
L-Theanine	20mg
Taurine	16.7mg
L-Phenylalanine	10mg
EVENING CELL SUPPLEMENT RECOMMENDATIONS	
Red algae extract	60mg
Cordyceps sinensis mushroom extract	3.3mg
L-Glutamine	100mg
L-Theanine	40mg
5-HTP	10mg
Jujube seed extract	1mg
Taurine	10.7mg

TABLE 5.2. CELL REGULATION COMPLEX	
L-Arginine	80mg
L-Ornithine	80mg
Herbal sleep blend: passion flower, lemon balm, chamomile flower, and poppy seed	17 mg

DNA REPAIR COMPLEX

When you strengthen your cells' DNA repair, you're giving your cells—and your entire body—a giant health boost. There's no better way to beef up this cell defense than with natural phytonutrients. Like broccoli, kale, and the other cruciferous veggies, mustard green extract will help your cells regulate and repair their DNA with ease. The natural antioxidant N-acetylcysteine will strengthen your free-radical defenses. When you finish your day with our recommended evening nutrients, you'll be getting all your daytime DNA repair boosters—plus melatonin, a natural hormone that promotes sound sleep while it protects your DNA from free-radical damage. Talk about a healthy nightcap.

TABLE 5.3. DNA REPAIR COMPLEX	
MORNING CELL SUPPLEMENT RECOMMENDATIONS	
Inositol hexaphosphate	10mg
Mustard green extract	0.8mg
Quercetin dihydrate	12mg
EVENING CELL SUPPLEMENT RECOMMENDATIONS	
Melatonin	0.8mg

TABLE 5.3. DNA REPAIR COMPLEX	
Inositol hexaphosphate	10mg
Mustard green extract	0.8mg
Quercetin dihydrate	12 mg

FREE RADICAL SCAVENGER COMPLEX

These days, it seems like all we do is fight toxicity with even more toxicity. That's why we love all-natural, toxin-fighting phytonutrients. Flavonoids—free-radical scavengers such as those found in bilberry fruit and grape skin—will destroy free radicals before they hit your cells. Meanwhile, ginkgo biloba leaf extract improves your blood flow as it fights off toxicity. Green barley grass contains a powerful antioxidant that can combat cancer, inflammation, cardiovascular diseases, and other conditions triggered by those pesky free radicals. Our evening recommendations include soothing rosemary and other gentle sedatives that will lull you to sleep.

TABLE 5.4. FREE RADICAL SCAVENGER COMPLEX	
MORNING CELL SUPPLEMENT RECOMMENDATIONS	
Bilberry fruit and grape skin extracts	2mg
Ginkgo biloba leaf extract	20mg
Green barley grass	10mg
EVENING CELL SUPPLEMENT RECOMMENDATIONS	
Cruciferous vegetable concentrate extract (broccoli, kale, radish)	16mg
Grape skin extract	8mg

TABLE 5.4. FREE RADICAL SCAVENGER COMPLEX	
Tomato lycopene extract	0.8mg
Rosemary extract	1.32mg
Pine bark extract	0.66mg
Lutein	0.33mg
Green barley grass	4 mg

STEM CELL MAINTENANCE COMPLEX

If there's one thing in your body that symbolizes youth, it's healthy stem cells. That's why the stem cell maintenance complex is such an important part of your anti-aging health routine. This complex's trio of algae will make sure your stem cells release quickly into your bloodstream right when you need them most. We've rounded off our stem cell health recommendations with two types of seaweed, which are ideal sources of natural antioxidants that work in synergy with the algae in this mix. Your stem cells will feel better than ever, and so will you.

TABLE 5.5. STEM CELL MAINTENANCE COMPLEX	
MORNING CELL SUPPLEMENT RECOMMENDATIONS	
Chlorella algae	10mg
Spirulina algae	20mg
Klamath blue-green algae	30mg
Fucoxanthin seaweed	10mg
Brown seaweed extract	0.8mg
EVENING CELL SUPPLEMENT RECOMMENDATIONS	
Chlorella algae	10mg

TABLE 5.5. STEM CELL MAINTENANCE COMPLEX	
Spirulina algae	20mg
Klamath blue-green algae	18mg
Fucoxanthin seaweed	10mg
Brown seaweed extract	2 mg

TELOMERE MAINTENANCE COMPLEX

A short telomere is a sign that your cells are aging. That's why we recommend a supplement plan full of phytonutrients that slow down premature telomere shortening. One of our favorite phytonutrients in this bunch is turmeric root extract, which cuts down on inflammation and minimizes some of the processes that shorten your telomeres. Fenugreek seed may be an ingredient in your favorite Middle Eastern dish, but it's also a natural way to regulate the length of your telomeres. When you add ginseng root and cayenne pepper fruit to the mix, your cells get an even bigger youthful boost. And speaking of boost, we've recommended milk thistle seed to help increase the quality of your z's so you're rested, energized, and ready to seize the day when morning rolls around.

TABLE 5.6. TELOMERE MAINTENANCE COMPLEX	
MORNING CELL SUPPLEMENT RECOMMENDATIONS	
Telomere Maintenance Complex	
Purslane extract	2mg
Turmeric root extract	10mg
Cayenne pepper fruit	4mg

TABLE 5.6. TELOMERE MAINTENANCE COMPLEX	
L-Arginine HCl	60mcg
Fenugreek seed	10mg
Korean ginseng root extract	20mg
Astragalus root extract	4mg
Borage seed oil	80mg
EVENING CELL SUPPLEMENT RECOMMENDATIONS	
Purslane extract	2mg
Turmeric root extract	10mg
Cayenne pepper fruit	4mg
Vanadium	8mcg
Fenugreek seed	8mg
Milk thistle seed extract	12mg
Borage seed oil	80 mg

DIGESTION ENHANCERS

With results this impressive, it would be a shame to run into absorption problems, which all the vitamin takers out there know is a major issue. That's why we recommend pairing these six phytonutrient complexes with digestive enzymes. These natural miracle workers will break down your phytonutrients—and all the other food you eat during the day—to ensure your body absorbs every drop of the health benefits.

Our recommended digestive enhancement complexes include all-natural digestive enzymes and helpful bacteria to help you absorb nutrients from your meals as well as your new phytonutrient friends. With root extracts to enhance vitamin absorption and enzymes to break down cellulose (insoluble fiber), lactose (milk sugar), and lipids (fats), this suggested mix will ease your digestion in many ways.

TABLE 5.7. DIGESTION ENHANCERS	
MORNING CELL SUPPLEMENT RECOMMENDATIONS	
Bacillus coagulans	120 million CFUs*
Black pepper fruit extract	2mg
Astragalus membranaceus root and Panax notoginseng root	10mg
Emblica officinalis fruit, Terminalia chebula fruit, Terminalia belerica fruit	10mg
Artichoke leaf extract	20mg
Guarana seed extract	16mg
Amylase	240 units**
Neutral protease	60 units
Cellulase	2 units
MORNING CELL SUPPLEMENT RECOMMENDATIONS (CONT.)	
Lactase	40 units
Lipase	10 units
EVENING CELL SUPPLEMENT RECOMMENDATIONS	
Bacillus coagulans	120 million CFUs
Black pepper fruit extract	2mg
Astragalus membranaceus root and Panax notoginseng root	10mg
Emblica officinalis fruit, Terminalia chebula fruit, Terminalia belerica fruit	10mg
Artichoke leaf extract	20mg

TABLE 5.7. DIGESTION ENHANCERS	
Amylase	160 units
Neutral protease	40 units
Cellulase	1.3 units
Lactase	26.7 units
Lipase	16.4 units
*CFUs = colony-forming units **Units = units of enzyme activity	

VITAMINS AND MINERALS

All this talk about phytonutrients doesn't mean that your trusty vitamins and minerals aren't important. By combining these vitamins and minerals with high-impact phytonutrients and digestive enzymes, you'll get the most out of them—plus all the cell-health benefits we've outlined. Get ready to feel a whole new enthusiasm for the vitamins and minerals you know and love.

TABLE 5.8. VITAMINS AND MINERALS	
MORNING CELL SUPPLEMENT RECOMMENDATIONS	
Vitamins	
Vitamin A	2,500IU
Vitamin C	40mg
Vitamin D	200IU
Vitamin E	15IU
Vitamin K	40mcg
Thiamin	2mg
Riboflavin	1.6mg

TABLE 5.8. VITAMINS AND MINERALS	
Niacin	28mg
Vitamin B6	8mg
Folate	20mcg
Vitamin B12	32mcg
Biotin	20mcg
Pantothenic acid	5mg
Minerals	
Calcium	120mg
Iodine	12mcg
Zinc	1.98mg
Selenium	12mcg
Copper	0.08mg
Manganese	0.08mg
Chromium	20mcg
Molybdenum	4mcg
EVENING CELL SUPPLEMENT RECOMMENDATIONS	
Vitamins	
Vitamin A	2,500IU
Vitamin C	80mg
Vitamin D	200IU
Vitamin E	15IU
Vitamin K	40mcg
Thiamin	1mg
Riboflavin	2mg
Niacin	28.2mg
Vitamin B6	3mg
Folate	32mcg

TABLE 5.8. VITAMINS AND MINERALS	
Vitamin B12	48mcg
Biotin	16mcg
Pantothenic acid	8mg
Minerals	
Calcium	120mg
Iodine	6mcg
Magnesium	60mg
Zinc	0.9mg
Selenium	9.6mcg
Copper	0.04mg
Manganese	0.04mg
Chromium	16mcg
Molybdenum	3.2 mcg

HOW YOU'LL FEEL

When we started our cell-health journey, one of the first things we noticed was an unbelievable boost in energy that lasted all day long. Our moods improved, and we were finally finding the drive to really embrace the phrase *carpe diem.* And that's just the immediate boost from our new supplement routines. Once we began embracing the interval exercise and the meditation, we knew there was no going back. We felt renewed from the inside out—and it's all thanks to the trillions of healthier cells in our bodies. We never realized how lousy we were feeling until we started feeling this good!

One of the best things about the Cell Solution lifestyle is that you don't have to wait months, or even weeks, to feel the difference. The difference starts today. When you dive into our recommended daytime nutrients, you will immediately get that youthful energy that lasts all day. Our suggested evening nutrients feature mild sedatives that promote the sound night's sleep you deserve. You'll be more alert during the day, more restful at night, and ready to apply that new vigor and energy to everything life throws your way. Meanwhile, your cells will be stronger, healthier, and ready to keep your body functioning at an all-time high. Who would say no to that?

Once you feel the benefits of boosting your cell health, you'll never want to look back. Rather than dreading the years to come, you'll be embracing each day with a new, healthy outlook. And while we're talking about those years down the road, we have one more youthful strategy to share. With the Cell Solution lifestyle, you've taken the time to nurture your cells so that they nurture you today. But what if you could take your newly healthy cells and save them for ten, or even twenty, years from now? You already have your youthful today. It's time to meet your youthful tomorrow.

THE TAKEAWAYS

- The Cell Solution supplement recommendations provide your cells with the perfect combination of all-natural phytonutrients, vitamins, minerals, and digestive enzymes.

- Take our morning and evening nutrient suggestions daily for the best anti-aging boost.

- Once you dive into our recommended supplements, you'll feel more energetic during the day and more rested at night.

- Combine these all-natural phytonutrients with cell-healthy diet, exercise, and meditation strategies to experience your youthful today.

COMING UP

- Discover how to take your stem cell health to the next level.

- Learn how bioinsurance is changing the way we think about aging.

EXPLORE MORE

- Visit www.youngertodaybook.com to learn more about these phytonutrients in action.

6

The Cell Solution Future

If you think suspending aging is science fiction, think again. We can achieve a more youthful future today.

We're no longer stuck with our genes or destined for decades of low energy and poor health. Those healthy sixties, seventies, and beyond are more in reach than ever before. Forget Leonardo da Vinci and Michelangelo; we're living in an anti-aging Renaissance, and stem cells are ushering it in.

We've talked a little about how nutrients and your daily habits can help keep your stem cells healthy and your genes working in your favor. Let's take that a step further. Around the world, and even in your neck of the woods, people are getting their adult stem cells collected so they can get a dose of health in the near future. Stem cell collection isn't science fiction, and it doesn't involve those controversial cells found in embryos. Stem cell collection is a safe, proven way to ensure you're healthy for as long as possible. By having those cells stored, you're giving yourself a powerful tool for fighting diseases and other illnesses down the road. Think of it as an all-natural body regenerative treatment for that unhealthy rainy day.

BIOINSURANCE: SCIENCE FUTURE, NOT SCIENCE FICTION

The term "life insurance" is a bit misleading. After all, that pricey policy doesn't really benefit you until you're, well, not alive. But what if we told you there's a type of insurance out there that's guaranteed to increase your quality of life while you're still around to reap the benefits? It all comes back to your stem cells.

"Bioinsurance" is a fancy term for stem cell collection, the process of removing your adult stem cells, freezing them, and saving them for future use. Here's how it works: you go to a stem cell collection facility. You're given a hormone that tells the healthiest, *crème de la crème* of stem cells to release from your bone marrow and enter your bloodstream. Then, your blood is drawn the same way as when you give blood—and it doesn't hurt one bit. Your blood is filtered so that only a high-quality stem cell concentration is left behind, and those stem cells are frozen at a stem cell bank until your physician decides you could use an influx of young, healthy cells. That time could come tomorrow, ten years from now, or beyond; it all depends on when your aging cells start outnumbering your healthy ones, and when aging diseases start developing. Once some of your healthy adult stem cells are put back in your body, you can reap all the benefits of their cell-copying powers. Soon, your healthy cell quota will go up and the toxic quota will go down. You'll feel reinvigorated, strong, and back to prime health.

If you think suspending aging is science fiction, think again. We can achieve a more youthful future today.

The best time to get your cells collected is between the ages of eighteen and forty, when your cells typically haven't started to turn toxic. But there's good news if you're older, too: collecting your stem cells at any age is better than not doing it at all. Your thirty-five-year-old cells are more fit than your forty-five-year-old ones. Even at fifty-five, there's still plenty of benefit to getting those stem cells collected and frozen before they continue their march toward toxicity. But it's not just a matter of getting that youthful boost. We're on the brink of discovering how collected stem cells can solve many important health issues. Want to know more? We thought you'd never ask.

MORPHING MAGIC: STEM CELLS IN USE TODAY

Stem cells are the best chameleons your body's ever known. They have the special ability to make healthy copies of your dying cells, and that's no small feat. Every day, we're exploring new ways to make the most of their talent, and it turns out that these little cells might be able to help us fight disease in big ways. Here's what we know so far:

To say heart disease is a big deal is the understatement of the century. It's the number-one cause of death in the United States, is claiming more and more lives each day, and causes major issues for people lucky enough to survive it. We all probably know someone affected by heart disease, and we'd rather not see another person live with that reduced quality of life. It's no surprise that scientists have been clamoring for a better way to treat heart disease—and their eyes are focused on those tiny stem cells. Your stem cells are great copiers, but what does that have to do with treating heart disease? A lot. Researchers are currently trying to determine whether we can apply stem cells' copying skills to the task of regenerating heart tissue. If it works, it could mean healthier, less damaged tissue for all those people who have suffered heart attacks. It's a pretty exciting possibility.

Lately, there's been a lot of talk about another way to apply stem cells' morphing magic: retina diseases. Scientists are finding more and more evidence that we can generate healthy retina cells using stem cells collected from our own bodies. These stem cells would replace nonfunctioning retina cells, which are the ones that often cause blindness or significant eyesight loss for people with retina diseases. So far, we've been able to improve the eyesight in animals with retina conditions through some strategic use of stem cells. Consider that a step in the right direction.

If all this disease fighting doesn't get you excited, we have one more major piece of news to share. We're starting to restore or reboot aging stem cells so that they function like younger, healthier cells. Although this is just now being done in labs, this is a big reason to consider stem cell collection even if you're beyond that eighteen-to-forty window. Imagine the difference it would make in your energy, health, and quality of life if you could get back your own cells today? And better yet, what if those cells were from your thirties, when you were at the prime of your health? This is in the works today. If you get your stem cells collected, you'll have the opportunity to take science up on this very youthful offer.

Your cell health future starts today, but that doesn't mean we shouldn't plan for a healthier tomorrow, too. Bioinsurance is the only life insurance policy you'll ever need.

A HEALTHIER TODAY

It doesn't matter whether we're talking about collecting stem cells or boosting them with nutritional complexes. The principle is the same: if our cells are healthy and youthful, so are we. Start with our diet, exercise, meditation, and nutrient strategies to get your cells as healthy as possible. Bioinsurance is the ultimate way to use your cells to your advantage, but even if you're not ready to go through that process quite yet, you can still do wonders for your cells by tapping into the Cell Solution lifestyle. The healthier your cells, the more beneficial your collection will be. So take those all-natural complexes that turn on your "good" genes and help minimize the "bad" ones. Exercise, eat, and de-stress your way to better cell health. See and feel the difference it makes when you boost your health from the cellular level. We have the resources to ensure a healthier, happier future. That "new you" is only a few cells away.

THE TAKEAWAYS

- Stem cell collection is a safe, painless way to ensure you'll have a boost of natural youthfulness when you need it most.
- Every day, researchers are finding more ways that stem cells can increase our health and ability to overcome diseases.
- Care for your stem cells now through our healthy cell lifestyle, and collect them at a registered stem cell facility to take their benefit to the next level.

COMING UP

- Recap the benefits of cell health with us.
- Get motivated to start your health journey today.

EXPLORE MORE

- Visit www.youngertodaybook.com for the latest news on stem cell research.

7

Your Younger Today

You're going through your hectic day, and somewhere along the way, your energy vanishes into thin air. You're dragging all afternoon. When you finally get home from work, exhausted, you happen to glance in the mirror as you walk to the kitchen. That's a mistake. Why do you look so fatigued? Is that a wrinkle coming in? Where's that vibrancy and zeal you had a few years ago? As you turn sideways to study your figure, a familiar, nagging thought pops into your mind. "If only I could feel young again."

Remember that familiar scenario? It's time to kiss it good-bye.

We're living in the golden age of anti-aging. After years of hearing confusing, conflicting health advice (how many more crash diets and exercise fads can we really take?) we finally know how to be truly healthy. And here's the best part: that healthier today doesn't require chemicals, gadgets, gimmicks, or spending our lives trapped at the gym. All it takes is a healthy cell lifestyle. With the diet, exercise, meditation, and supplement habits we've outlined, you can take control of your genes and age gracefully. All-natural supplements will give your cells all the nutrients they crave in a winning combo that triggers just about every health benefit in the book. It's juicing without the cleanup, or a trip to the produce market without the hassle of finding parking. It's "back to the basics" with a big boost, and it's within reach today.

We know you hear new health advice every day, but cell health is the real deal—and you'll know it the moment you start the plan. This isn't a "feel better in thirty days" regimen; it's a "feel better today" one. Talk about instant gratification! When we first started the plan, we went from being fatigued—and admittedly a little moody—to being active, upbeat, and a lot more fun to be around. Just ask anyone who knows us; the difference was huge, and it happened overnight.

With these phytonutrients on your side, your energized days will lead to restful nights. If you've had a bad night's sleep before, we don't have to describe to you how much it affects everything else you've got going on. It all comes back to that biorhythm! With our plan, we're getting our best sleep ever, and our days are feeling a lot more youthful because of it. We're feeling like the best version of ourselves, and we're looking a lot healthier, too. Once we paired our natural supplements with cell-healthy diet, exercise, and meditation habits, we couldn't wait to stand in front of the mirror. Our skin became firm, less wrinkled, and glowing; our bodies got a little slimmer; and needless to say, our smiles were absolutely contagious. They still are.

We've shown you the secret to more youthful aging. We've described why your health has to start and end with your cells. We've given you the diet, exercise, and meditation habits that lay a cell-healthy foundation, and topped that off with the nutrient recommendations that give your cells the biggest health boost. You have the plan and the resources, and you have our support to motivate you. Now it's time for action.

The future of aging has never been brighter. We've never known more about what it takes to stay healthy, and new research is making that elusive fountain of youth a little more within reach every day. Organizations like the Cell Health Institute are leading the charge. By supporting stem cell research and developing natural products that make groundbreaking anti-aging science attainable for all of us, the Cell Health Institute is helping us discover your younger today. We're getting closer to understanding the real potential of stem cells in our quest for youthfulness, and believe us, this isn't lab research that sits in file folders collecting dust for decades. These breakthroughs will be real game changers, and it'll happen sooner than you think. Our stem cells could help us recover from heart disease and other conditions that we've seen so many people suffer through. Those stem cells can unlock our youthful future if we make them a priority in our daily and long-term health routines. Your tiny stem cells can be the difference between spending your retirement years at the doctor's office or embracing the world around you. We know which future we're hoping for!

It doesn't matter how you dive into the healthy cell lifestyle; the most important thing is to simply do it.

Whether you foster your anti-aging through our supplement and lifestyle suggestions or take it a step further with stem cell collection, your healthier tomorrow can start today. It doesn't matter how you dive into a healthy cell lifestyle; the most important thing is to simply do it. Don't wait five years, or even five days, to give yourself the health and well-being you deserve. The sooner you start nurturing your cells, the sooner you may achieve all those things you've been dreaming about. Travel the world. Have more energy for your passion projects. Spend more quality time with your family. Dive into new hobbies. And, at the very least, wake up feeling rested and happy. It's a quality of life we all deserve, and for the first time in our lives, we have the resources to make it happen. What are you waiting for?

There's no more hiding from the mirror. There's no more looking back on your week and wondering why you didn't have the energy to accomplish more. And there's no more resigning yourself to years of disease, unhappiness, and a poor quality of life. We've given you the information that changes your habits, and we hope you'll use it to develop the habits that transform your life. Aging is a fact of life, but it's also largely in our control. We're blasting off into our most promising, healthy years yet, and it starts with a youthful today.

That's what we see for our future. What do you see for yours?

Appendix

A Quick Guide to Your Healthy Cell Day

Want to take your healthy, youthful lifestyle with you wherever you go? Your wish is our command. Here's the quick guide to your healthy cell day.

EXERCISE FOR CELL HEALTH

Get your cells moving in a more youthful direction.

EXERCISE GOALS: Make it your goal to improve body composition, not the number on the scale. When you increase muscle mass and decrease body fat, your hormone levels will improve, your pH levels will balance, and you'll have more strength in your daily activities.

MORNING EXERCISE: Exercise lightly for fifteen to twenty minutes each morning before breakfast so that you'll burn more calories for the rest of the day.

CORE EXERCISE: Exercise your lower back and abdomen every day to avoid lower back pain, poor posture, disc injuries, and other signs of aging.

CARDIO EXERCISE: Exercise in cycles. Push yourself to two-thirds of your max for two minutes, then pause for a few minutes to give your cells the rest they need. Not sure what exercises to do? Try jumping jacks or hand, vest, or leg weights

STRETCHING EXERCISES: Practice these five stretches daily to benefit both your body and mind.

EAT FOR CELL HEALTH

Eat your way back to health.

MEALS: Make breakfast the biggest meal of your day. Lunch should be a little smaller, and dinner should be the smallest meal of all. Your diet should be divided almost equally among protein, carbs, and fat. We recommend the 40-30-30 ratio.

PROTEIN: Always eat protein first, and choose poultry, fish, and tofu over red meat.

FATS: Don't cut out "good" fats like omega-3s. Avoid high-fat sauces and creams that add many calories but not much nutritional value.

SNACKS: Eat two per day between meals to keep your blood sugar steady. Some of our favorite snacks include low-glycemic fruit such as apples, pears, and figs, and proteins such as cottage cheese, string cheese, tofu, and edamame. When you're choosing a protein bar, select one that offers a 40-30-30 or 40-20-20 balance of carbohydrates, protein, and fats. If you're hungry before bed, pair fruit with protein for a healthful snack.

PORTIONS: At each sitting, only eat as much food as can fit in the palm of your hand. Chew your food thoroughly, and drink a glass of water mixed with two tablespoons of olive oil before meals to cut your cravings.

LIQUIDS: Hydration is essential! Drink at least eight 8-ounce glasses of water each day, and even more if you're exercising regularly or perspiring heavily due to warm climate. It's also important to have your water as close to alkaline as possible, and there are plenty of products on the market that can be dropped into your water to increase its pH to this optimal state.

SUPPLEMENTS: Follow the cell-health formula in Chapter 6 to give your cells a boost, morning and night.

DE-STRESS FOR CELL HEALTH

Send your stress to the hills.

BREATHING: When you can't seem to relax, take a long, deep breath. Breathe in and out, with the entire cycle lasting about one minute. Repeat this four more times, focusing on inhaling and exhaling deeply. Do this breathing exercise three times per day to keep your stress at bay.

MEDITATION: Having trouble living in the moment? This simple meditation will help. Close your eyes and visualize a field and a mouse hole. Envision each of your distracting thoughts as mice that go into the mouse hole, where they are out of sight and no longer occupying your mind. Once you can count twenty seconds between seeing any mice, you're thinking in the moment. As an added bonus, those cortisol levels have likely dropped, as well.

MANTRA: Every night before you go to sleep, utter the same phrase. Our suggestion: "The Cell Solution is my source of information for better cell health—now and in the future." This trigger phrase will tell your cells that it's time to start their nightly repairs.

Selected References

This abridged list includes some of the scientific studies that support the *Younger Today* approach to health. Please visit www.youngertodaybook.com/ **references for the complete list.**

Addis, Russell C. and Jonathan A. Epstein. "Induced Regeneration—the Progress and Promise of Direct Reprogramming for Heart Repair." *Nature Medicine* 19(July 2013): 829–836.

Aguirre, Aitor, Ignacio Sancho-Martinez, and Juan Carlos Izpisua Belmonte. "Reprogramming Toward Heart Regeneration: Stem Cells and Beyond." *Cell Stem Cell Review* 12(March 2013): 275–284.

Arias, Alberta J., Karen Steinberg, and Robert L. Trestman. "Systematic Review of the Efficacy of Meditation Techniques as Treatments for Medical Illness." *The Journal of Alternative and Complementary Medicine* 12(2006) 817–832.

Atkinson, Greg and Damien Davenne. "Relationships Between Sleep, Physical Activity and Human Health." *Physiological Behavior* 90(February 2007): 229–235.

Bachstetter, Adam D., Jennifer Jernberg, Andrea Schlunk, et al. "Spirulina Promotes Stem Cell Genesis and Protects Against LPS Induced Declines in Neural Stem Cell Proliferation." *PLos ONE* 5(2010).

Bkasin, Manoj K., Jeffrey A. Dusek, Bei-Hung Chang, et al. "Relaxation Response Induces Temporal Transcriptome Changes in Energy Metabolism, Insulin Secretion and Inflammatory Pathways." *PLoS ONE* 8(2013).

Choi, Sang-Woon and Simetta Friso. "Epigenetics: A New Bridge Between Nutrition and Health." *Advances in Nutrition* 1(2010): 8–16.

Ciolac, Emmanuel Gomez. "High-intensity Interval Training and Hypertension: Maximizing the Benefits of Exercise?" *American Journal of Cardiovascular Disease* 2(2012): 102–110.

Delezie, Julien and Etienne Challet. "Interactions Between Metabolism and Circadian Clocks: Reciprocal Disturbances." *Annals of the New York Academy of Sciences* 1243(2011): 30–46.

Denham, Joshua, Francine Z. Marques, Brendan J. O'Brien, et. al. "Exercise: Putting Action into our Epigenome." *Sports Medicine* (October 2013).

Dusek, Jeffery A., Hasan H. Out, Ann L. Wohlhueter, et al. "Genomic Counter-Stress Changes Induced by the Relaxation Response." *PLoS ONE* 3(2008).

Drust, B., J. Waterhouse, G. Atkinson, et al. "Circadian Rhythms in Sports Performance—An Update." *Chronobiology International* 22(2005): 21–44.

Eckel-Mahan, Kristin and Paolo Sassone-Corsi. "Metabolism and the Circadian Clock Converge." *Physiological Review* 93(2013): 107–135.

Ekmekcioglu, C. and Y. Touitou. "Chronobiological Aspects of Food Intake and Metabolism and their Relevance on Energy Balance and Weight Regulation." *Obesity Reviews* 12(2010): 14–25.

Eksteins, Angela. "Meditation May Be the Future of Anti-Aging, Part II." *Natural News* (February 2010).

Esquirol, Yolande, Vanina Bongard, Laurence Mabile, et al. "Shift Word and Metabolic Syndrome: Respective Impacts of Job Strain, Physical Activity, and Dietary Rhythms." *Chronobiology International* 26(2009): 544–559.

Fitton, Janet Helen. "Therapies from Fucoidan: Multifunctional Marine Polymers." *Marine Drugs* 9(2011): 1731–1760.

Frei, Balz and Jane V. Higdon. "Antioxidant Activity of Tea Polyphenols in Vivo: Evidence From Animal Studies." *The Journal of Nutrition* 133 (2003): 2375S–3284S.

Gale, John E., Heather I. Cox, Jingy Qian, et al. "Disruption of Circadian Rhythms Accelerates Development of Diabetes through Pancreatic Betal-Cell Loss and Dysfunction." *Journal of Biological Rhythms* 26(2011): 423–433.

Giampapa, Vincent C., Frederick F. Buechel, and Ohan Karatoprak. *The Gene Makeover: The 21st Century Anti-Aging Breakthrough.* Laguna Beach, CA: Basic Health Publications, Inc., 2007.

Gibala, Martin J., Jonathan P. Little, Maureen J. MacDonald, et al. "Physiological Adaptations to Low-Volume, High-Intensity Interval Training in Health and Disease." *The Journal of Physiology* 590.5(2012): 1077–1084.

Hafstad, A.D., N.T. Boardman, J. Lund, et al. "High Intensity Interval Training Alters Substrate Utilization and Reduces Oxygen Consumption in the Heart." *Journal of Applied Physiology* 111(2011): 1235–1241.

Hall, Martica H., Michele L. Okun, MaryFran Sowers, et al. "Sleep is Associated with the Metabolic Syndrome in a Multi-Ethnic Cohort of Midlife Women: The SWAN Sleep Study." *Sleep* 35(2012): 783–790.

Hastings, Michael H., Akhilesh B. Reddy, and Elizabeth S. Maywood. "A Clockwork Web: Circadian Timing in Brain and Periphery, in Health and Disease." *Nature Reviews* 4(August 2003): 649–661.

Jacobs, Tonya L., Elissa S. Epel, Jue Lin, et al. "Intense Meditation Training, Immune Cell Telomerase Activity, and Psychological Mediators." *Psychoneuroendocrinology* 36(2011): 664–681.

Jensen, G.S., A.N. Hart, L.A. Zaske, et al. "Mobilization of Human CD34+CD133+ and CD34+CD133(-) Stem Cells in Vivo by Consumption of an Extract from

Aphanizomenon Flos-Aqua-Related to Modulation of CXCR4 Expression by an L-Selectin Ligand." *Cardiovascular Revascularization Medicine* 8(2007): 189–202.

Lila, Mary Ann. "Anthocyanins and Human Health: An In Vitro Investigative Approach." *Journal of Biomedicine and Biotechnology* 5(2004): 306–313.

Liou, C.H., C.W. Hsieh, D.Y. Chen, et al. "Detection of Nighttime Melatonin Level in Chinese Original Quiet Setting." *Journal of the Formosan Medical Association* 109(October 2010): 694–701.

Niculescu, Mihai D. and Daniel S. Lupu. "Nutritional Influence on Epigenetics and Effects on Longevity." *Current Opinion in Clinical Nutrition and Metabolic Care* 14(2011): 35–40.

Ong, J.M. and L. da Cruz. "A Review and Update on the Current Status of Stem Cell Therapy and the Retina." *British Medical Bulletin* 102(May 2012): 133–146.

Patel, Sonal, Nikkhil Velingkaar, Roman V. Kondratov, et al. "Transcriptional Control of Antioxidant Defense by the Circadian Clock." *Antioxidants and Redox Signaling* (October 2013).

Ramsden, C.M., M.B. Powner, Amanda-Jayne F. Carr, et al. "Stem Cells in Retinal Regeneration: Past, Present and Future." *Development* 140 (2013): 2576–2585.

Ribaric, Samo. "Diet and Aging." *Oxidative Medicine and Cellular Longevity* (2012): 1–20.

Ronn, T., P. Volkov, C. Davegardh, et al. "A Six Months Exercise Intervention Influences Genomewide DNA Methylation Pattern in Human Adipose Tissue." *PLOS Genetics* 9(June 2013): 1–16.

Saatcioglu, Fahri. "Regulation of Gene Expression by Yoga, Meditation and Related Practices: A Review of Recent Studies." *Asian Journal of Psychiatry* 6(2013): 74–77.

Sanganalmath, Santosh K. and Roberto Bolli. "Cell Therapy for Heart Failure: A Comprehensive Overview of Experimental and Clinical Studies, Current Challenges, and Future Directions." *Circulation Research* 113(August 2013): 810–834.

Schulman, Ivonne H. and Joshua M. Hare. "Key Developments in Stem Cell Therapy in Cardiology." *Regenerative Medicine* 7(November 2012): 17–24.

Shiraev, Tim. "Evidence Based Exercise: Clinical Benefits of High Intensity Interval Training." *Australian Family Physician* 41(December 2012): 960–962.

Tammen, Stephanie A., Simonetta Friso, and Sang-Woon Choi. "Epigenetics: The Link Between Nature and Nurture." *Molecular Aspects of Medicine* 34(2013): 753–764.

Tooley, G.A., S.M. Armstrong, T.R. Norman, et al. "Acute Increases in Nigh-Time Plasma Melatonin Levels

Following a Period of Meditation." *Biological Psychology* 53(2000): 69–78.

Viczian, Andrea S. "Advances in Retinal Stem Cell Biology." *Journal of Ophthalmic and Vision Research* 8(2): 147–159.

Viuda-Martos, M., Y. Ruiz-Navajas, J. Fernandez-Lopez, et al. "Functional Properties of Honey, Propolis, and Royal Jelly." *Journal of Food Science* 73(2008): 117–124.

Winter, Christopher W., William R. Hammond, Noah H. Green, et al. "Measuring Circadian Advantage in Major League Baseball: A 10-Year Retrospective Study." *International Journal of Sports Physiology and Performance* 4(2009): 394–401.

About the Authors

Vincent C. Giampapa, M.D., F.A.C.S., is a leading global expert on anti-aging medicine. A board certified plastic and reconstructive surgeon, he has spent decades advancing research on aging so that his patients and others can achieve a healthier, more youthful life. "Dr. G" has shared his expertise in countless publications and six books, including *The Gene Makeover: The 21st Century Anti-Aging Breakthrough.* He was a founder of the American Academy of Anti-Aging Medicine, the first president of the American Board of Anti-Aging Medicine, and one of the world's first certified anti-aging medicine physicians. Dr. G continues to bring anti-aging advancements to the public as the Chief Medical Officer and co-founder of the Cell Health Institute, and Clinical Professor at Rutgers University. He recently presented his breakthrough research on stem cells and anti-aging at U.S., European, and Scandinavian health conferences.

Carol Alt is a supermodel and raw foods expert who has inspired countless people to adopt healthier, more natural lifestyles. She is the author of three bestselling noncooked lifestyle books, and shares her nutritional knowledge with millions of viewers per week as host of the FOX News Channel program "A Healthy You & Carol Alt." Through her three decades as an internationally celebrated model, Carol has graced the cover of more than 700 magazines and counting. Her resume also includes TV host, philanthropist, and award-winning actor.

Made in the USA
Monee, IL
07 May 2026

49715371R10070